PROJECT MANAGEMENT OF COMPLEX AND EMBEDDED SYSTEMS

Ensuring Product Integrity and Program Quality

Kim H. Pries
Jon M. Quigley

CRC Press
Taylor & Francis Group
Boca Raton London New York

CRC Press is an imprint of the
Taylor & Francis Group, an **informa** business

AN AUERBACH BOOK

CRC Press
Taylor & Francis Group
6000 Broken Sound Parkway NW, Suite 300
Boca Raton, FL 33487-2742

First issued in paperback 2019

© 2009 by Taylor & Francis Group, LLC
CRC Press is an imprint of Taylor & Francis Group, an Informa business

No claim to original U.S. Government works

ISBN-13: 978-1-4200-7205-1 (hbk)
ISBN-13: 978-0-367-38662-7 (pbk)

Library of Congress Cataloging-in-Publication Data

Pries, Kim H., 1955-
  Project management of complex and embedded systems : ensuring product integrity and program quality / authors, Kim H. Pries, Jon M. Quigley.
    p. cm.
  Includes bibliographical references and index.
  ISBN 978-1-4200-7205-1 (hardcover : alk. paper)
  1. Project management. 2. Quality control. I. Quigley, Jon M. II. Title.

HD69.P75P673 2009
658.4'013--dc22   2008038297

Visit the Taylor & Francis Web site at
http://www.taylorandfrancis.com

and the CRC Press Web site at
http://www.crcpress.com

Contents

List of Figures

List of Tables

Preface

This book is written to guide the project manager or project participant through the development process. We understand and have used other project management approaches and models. This book is designed to provide information and guidance about the automotive approach to the following constituencies:

- Automotive project/program managers
- Project/program managers in other industries requiring many controls on the process (food industry or airline industry)
- Service industries such as hospitality and hospitals
- Embedded teams looking for control
- Organizations that certify to ISO/TS 16949:2002
- Organizations that certify to ISO 9001:2000
- Medium-to-heavy manufacturing companies with project management in their armamentaria
- Universities training engineers and other students for careers in industry

We include some information derived from Department of Defense (DoD) sources because the U.S. defense industry is really the root source for program and project management skills. Even some of the apparently automotive tools originated with the DoD; for example, the failure mode and effects analysis (FMEA) derived from MIL-STD-1629A, which is failure mode, effects, and criticality analysis. A large portion of the DoD material is still relevant today and still used by DoD project managers.

Other ideas are part of a system developed by General Motors, Chrysler, and Ford to standardize the approach to designing and developing new products. The Automotive Industry Action Group (AIAG) supplies a substantial amount of support to automotive techniques, particularly in the form of easy-to-understand manuals that describe such concepts as:

- Statistical process control (SPC)
- Measurement systems analysis (MSA)
- Advanced product quality planning (APQP)
- Failure mode and effects analysis (FMEA)
- Machinery failure mode and effects analysis (MFMEA)
- Quality system assessment (QSA)
- Quality management systems

In general, we have avoided extended discussions of areas we feel are more germane to a functioning quality management system and we have tried to keep our focus on the project management side of development. As far as we know, we are the first to delve more deeply into automotive project management.

Acknowledgments

Jon Quigley would like to thank Nancy Quigley and Jackson Quigley for all their patience and encouragement. I would also like to thank my parents for putting me in a position to be able to do the things I have been able to do.

He would also like to thank Subramanian Arumugam, who assisted with the embedded development overview, and Fred Starkey who provided the graphic for an example of product test flow. Thanks also to John Lyons, chief engineer of the electrical and electronics groups for Volvo Trucks 3P, and 11660 for their support and providing such a great place to learn.

Also, a thank you to all of the reviewers of the book listed below.

- Jim Gwynn
- Ray Gardea
- John Bate
- François Longueville

Kim Pries would like to thank Janise Pries for painstakingly reviewing the manuscript for language and grammar issues. Any remaining errors rest with Jon and me. I would also like to thank Mark Tervalon, president of Stoneridge Electronics and Mike Carnahan, vice-president/general manager of Stoneridge Electronics—North America, for their continued support. Jim Spiller of www.criticaltools.com provides much-needed plug-in tools for Microsoft Project®, enhancing an already powerful product.

Those who think our dissemination of the "automotive way" is heavy-handed should realize that many of these ideas are currently being introduced into a school system in Texas with some success.

About the Authors

Jon M. Quigley is the manager of the electrical/electronics systems and verification group for Volvo 3P in Greensboro, North Carolina. As the manager of the systems group, he is responsible for ensuring a technical solution that will meet the customer's quality and cost demands of the various vehicle platforms. The verification portion of the group is responsible for verifying the system solution of the appropriate functionality, quality and reliability. Jon was part of the team that won both the Volvo 3P Technical Recognition Award (2005) and the Volvo Technology Award (2006). He has been issued four patents with another four patents in various phases at the patent office. He has spent the majority of his career in embedded (software and hardware) development. Quigley holds a bachelor's degree in engineering from the University of North Carolina at Charlotte and two master's degrees, an MBA in marketing, and a Master of Science in project management from Seattle City University. He is a member of PMI and is a certified Project Management Professional (PMP). He is also a member of SAVE International (The Society of American Value Engineers).

Kim H. Pries is director of product integrity and reliability for Stoneridge Electronics North America (SRE-NA), a division of Stoneridge, Incorporated. As an executive for SRE-NA, he is responsible for the quality management system at five locations, the validation and calibration laboratory, reliability, document management and engineering change, software testing, and production test equipment. He holds two bachelor's degrees and two master's degrees—three of them in engineering and the last one from Carnegie-Mellon University. Additionally, he holds the Certified Production and Inventory Manager (CPIM) designation from APICS and the following from ASQ: Certified Quality Auditor (CQA), Certified Quality Engineer (CQE), Certified Six Sigma Black Belt (CSSBB), and Certified Reliability Engineer (CRE). During the defense contractor phase of his career, he served as proposal manager (project manager) for several defense proposals. He authored and is revising a previous book on achieving the Six Sigma certification. By using Six Sigma projects and project management, he has been part of an initiative saving SRE-NA millions of dollars per year in cost reductions, cost avoidances, and profit improvements.

Chapter 1

Projects and Project Managers

1.1 Delivery

1.1.1 Overview of Program/Project Management

Many of the examples we present throughout this book come from the automotive supplier and customer world because we work in that environment every day. We suggest that most of the automotive development and production ideas have universal value—particularly the concept of process and design controls. In some cases, we will identify where a particular approach conveniently satisfies embedded development or, in other cases, service process development.

We distinguish automotive project management from general project management because the International Organization for Standardization (ISO) spells out the documentation and action requirements in an international standard (ISO/TS 16949). The Production Part Approval Process (PPAP) alone generates a minimum of 18 documents as a standard part of the package. Also, the stakes involved in automotive product release use large quantities of invested capital and expenses—a new vehicle runs into the tens of millions of dollars and sometimes more.

The automotive superset of ISO 9001:2000, the ISO standard ISO/TS 16949, regulates automotive project management.

Whatever the standards, we can generalize the automotive approach to nearly any industry. In fact, a little research reveals that the Hazard Analysis and Critical Control Point (HACCP) standard used by the food industry works as a direct analog of the automotive process control sequence.

1.1.2 Need for This Book

Currently, no book exists that promotes and generalizes the automotive project and program management approach in detail. General project management texts supply generic information and construction-oriented works support those disciplines. While these works perform well as references, they do not particularly help the project manager who is involved in capital-intensive projects or who desires to implement the variety of controls derived from automotive-style development and implementation.

Our QP (Quigley Pries) model (Figure 1.1) promotes the idea of project and program management as a control system; that is, we find from experience that the best project management is that which is self-regulating. Furthermore, the automotive approach is full of "controls" that keep bad things from happening. We think the idea of "control" generalizes nicely to a kind of project management that regulates itself. In the graphic below, the planning activities and organizational processes/procedures define the feedback and control loops. The sample frequency (meetings and communications) and the key variables to control (metrics) are identified and ongoing project actions respond according to the system design.

Successful projects of any stripe rely on the age-old concepts of anticipation, execution, and follow-through. We will show project managers how the tools we use every day provide benefits and success while reducing the nuisances and bypassing struggles.

1.1.3 *Comparison of Approaches*

We will demonstrate how the various approaches to project management relate to each other. Not only does the *compare and contrast* section have pedagogical value, but it should also encourage cross-pollination among project management approaches. This interapproach exposition will occur throughout the book. We will also show—with the exception of the details—that the diverse approaches to program/project management behave as interpretations of the main theme of project management.

1.1.3.1 Automotive Industry Action Group

The Automotive Industry Action Group (AIAG) publishes a package of seven paperbound resources sometimes known as "the magnificent seven." These books define the following areas of automotive interest:

- Quality management systems
- Quality system assessment (QSA)
- Measurement system assessment (MSA)
- Production part approval process (PPAP)

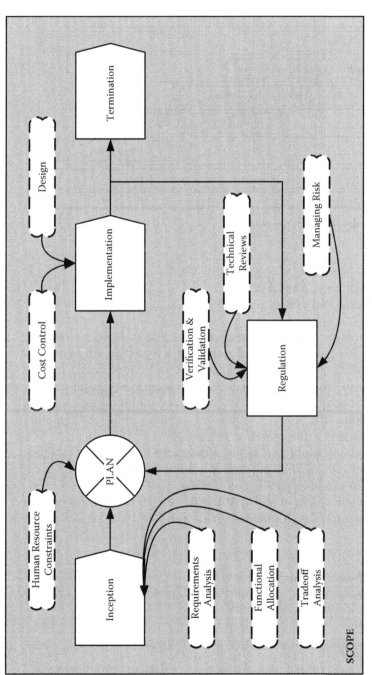

Figure 1.1   The QP model.

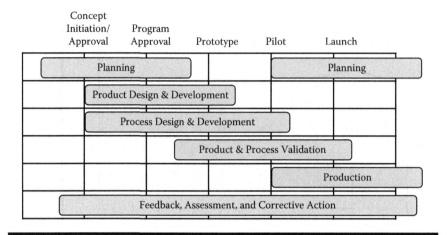

Figure 1.2   AIAG development process.

- Advanced product quality planning (APQP)
- Statistical process control (SPC)
- Failure mode and effects analysis (FMEA)

Figure 1.2 illustrates the project phases according to AIAG.[1] However, while this figure implies time dependencies, we selected an arbitrary horizontal representation of the phases for illustration only. Specifically, it is not always desirable to have the product design and process design happen simultaneously, although the team may choose concurrent engineering for competitive reasons. The team considers the process from the beginning without dominating or stifling product design.

The AIAG also sells numerous other publications in support of automotive design, development, and production. One version evolved as a variant of FMEA: the machinery FMEA or MFMEA.

Of the seven principal works by AIAG, the APQP publication defines automotive design and development for new products. The phases of the APQP are

1. Planning
2. Product design and development
3. Process design and development (manufacturing- or service-oriented)
4. Product and process validation
5. Production
6. Feedback assessment and corrective action (all phases)

APQP represents a useful template for program management. It presents a rational approach to product and process development and it generalizes to services and embedded development.

1.1.3.2 Department of Defense

The United States Department of Defense (DoD) approach to projects and programs uses significant milestones such as formal design reviews. MIL-STD-1521B defines these reviews to be:

1. System requirements review (SRR)
2. System design review (SDR)
3. Software specification review (SSR)
4. Preliminary design review (PDR)
5. Critical design review (CDR)
6. Test readiness review (TRR)
7. Functional configuration audit (FCA)
8. Physical configuration audit (PCA)
9. Formal qualification review (FQR)
10. Production readiness review (PRR)

Clearly, if the development team works on a software-free subsystem, the software reviews disappear from the project plan.

1.1.3.3 Institute of Electrical and Electronics Engineers

The principal Institute of Electrical and Electronics Engineers (IEEE) document defining project management is IEEE-1220, an updated and more detailed version of MIL-STD-499B (draft), a military standard that never became a full standard. The typical organization of a project under IEEE-1220 is

1. System definition
2. Subsystem definition
3. Preliminary design
4. Detailed design
5. Fabrication, assembly, integration, and test
6. Production
7. Customer support

1.1.3.4 Electronics Industry Association

Other standards for system development exist beyond IEEE-1220. These include Electronic Industries Alliance (EIA)-632, *Processes for Engineering a System* and ISO/IEC-15288, *Systems Engineering: System Life Cycle Processes* among others. Also, standards organizations seemed to have developed a penchant for further refining organization/process models into a new format called "Maturity Models." These models define an aging/increase in wisdom approach to organization improvement. EIA-632 looks at programs such as:

1. Assessment of opportunities
2. Investment decisions

3. Systems concept development
4. Subsystems design and predeployment
5. Deployments, operations, support, and disposal

1.1.4 Process Management Areas

The process management areas found in all project management approaches include many of the following:

- **Project management processes** – collecting management tasks
- **Requirements analysis** – defining the scope
- **Functional analysis and allocation** – further defining the scope, the divergence action
- **Design synthesis** – taking ideas and putting them together, the convergence action
- **Verification and validation** – ensuring we meet quality requirements
- **Process outputs** – each step in development of a product, service, or embedded software has deliverables
- **Work breakdown structure** – the heart of resource management and cost allocation
- **Configuration management** – controlling hardware and software release so that we know what we have
- **Technical reviews and audits** – keeping the program/project on track
- **Tradeoff analyses** – are we checking all of our options?
- **Modeling and simulation** – wonderful when we can do them
- **Metrics** – introducing objectivity
- **Risk management** – managing the inevitable challenges that arise
- **Planning** – schedule, cost, and quality
- **Product improvement strategies** – during development and after
- **Integrating system development** – putting all components together
- **Contractual considerations** – managing commercial issues
- **Management oversight** – part of any quality system

We feel that configuration management and product improvement strategies specifically belong to quality management systems like ISO 9001:2000. During the course of this book, we will emphasize the items we know, from our experience, belong to project and program management.

This book has illustrations of all the major operations that must occur for successful project management to happen. These graphics do not present every input, output, and interaction, but rather the key concepts necessary to implement a project.

1.1.5 Staged-Gate Methods in General

1.1.5.1 What Is a Gate?

A gate occurs as a milestone in a project schedule where the project team reviews the contents or deliverables of the phase at the end of that phase to decide if the project moves on to the next phase. Then, the team observes, reviews, measures, quantifies, evaluates, and critiques the project to determine if the project is at a point where designated gatekeepers decide to move to the next phase or reject the project. The team reviews the output deliverables as a check to ensure they meet the input requirements of the next phase. Once the team completes a phase gate, the next phase commences with no reversal to the previous phase—a one-way trip.

1.1.5.2 Reasons for Gate Reviews

We also call the gate review "kill-points." At the end of any particular phase, the team conducts a review to verify that project deliverables will fulfill the original needs defined by the development process. Frequently, the enterprise will wed the gate review with a business environment review as a means of verifying the relevance of the project. If the competitive landscape changed significantly, the program or project is no longer relevant and therefore merits discontinuation. Also, customers (internal and external) must know that gates function as kill-points and the program manager must not fear to inform the team that gates are kill-points.

Additionally, reviews allow for recalibration of the schedule. Even as the team evaluates actions that have taken place, they should consider the relevance of the existing project schedule. Difficulties occur during recalibration of the schedule if customers refuse to modify their schedules. Our experience suggests letting the customer become aware of schedule risk immediately rather than delay until the supplier or customer relationship degenerates.

Risk management generally includes the topics of management risk identification, planning, assessment, and handling (sometimes called "mitigation"), see Figure 1.3. We find an abundance of project management books that detail the techniques for handling risk. One such book, *Project & Program Risk Management: A Guide to Managing Project Risks and Opportunities*, by R. Max Wideman [Wideman 1992], is a good source for learning how to manage risk.

1.1.5.3 Objectives

If the reason for undertaking the project no longer remains valid or if the project output misses achieving project goals, then early awareness helps reduce the financial exposure of the project (that is, the amount that the enterprise loses on the project).

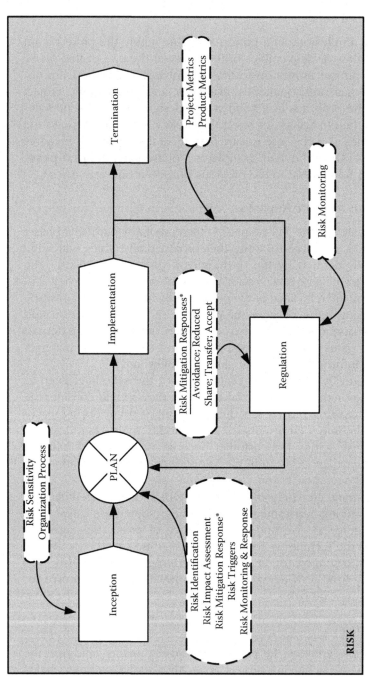

Figure 1.3 Risk management.

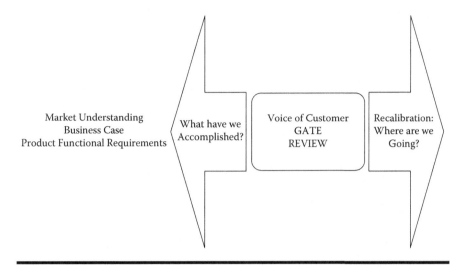

Figure 1.4   Gate reviews.

The goals of a review are Janus-like (see Figure 1.4) in that a review looks forward and backward. In the backward-looking portion of the review, the team, with the executive staff present, recapitulates the progress of the project to date and assesses the status of the project as it stands during the review. In the forward-looking portion of the review, the team assesses the reality of the schedule, performs risk analysis, and updates the budget. The team executes these reviews with a critical eye and a willingness to terminate the project or program if necessary or prudent.

The team conducts these reviews in order to

- Ensure on-track project deliverables (schedule/scope/cost),
- Compare deliveries to date with the reason for the project and terminate if necessary, and
- Ensure the next project phase identifies requisite inputs to promote success.

1.1.5.4   Gate Targets

The program manager, in concert with the project or launch team, selects gate targets to improve the probability of success and to limit the financial damage to the organization if the project becomes a failure—they function as decision points. Many companies have a published launch process with predefined gate targets; for example, pilot runs, run-at-rate (simulated full production), and start of production. Other companies may use in-process reviews scheduled on a regular basis (e.g., biweekly) to measure progress and assess risk. In some cases, a gate target fiasco drives the gatekeepers to terminate the project during or before the review.

Any factor that leads to a disruption of program or project progress becomes a candidate for risk assessment. These factors derive from either internal sources of the enterprise or arise from an external source (e.g., government regulatory requirements). Gate targets function as a control mechanism to certify that action occurs regarding project continuation or termination.

For early gate targets, the team will want to test critical and high-risk product design features and test new manufacturing process techniques. Some gate issues revolve around mandatory industry certifications; for example, PCS Type Certification Review Board (PTCRB) for *Groupe Spécial Mobile* or Global System for Mobile communications (GSM) wireless devices.

1.1.5.5 Importance of Gate Target Selection

The gate target approach lends structure to the project. Each segment of the process leading up to a gate target will have cost, quality, and schedule goals built into it. Frequently, program managers will define the gate targets in terms of entrance and exit criteria and record the results in a simple checklist. Where possible, the project manager defines the decision path on key gate targets before a project starts or well before a review is done on a specific target.

1.1.5.6 Measurement of Targets (Statistics)

Keeping measurement on target dates and target spending favors future project managers by allowing for statistical analysis based on project histories. For a given project manager, each kind of scheduled activity begins to take on a characteristic probability distribution. In some cases—for example, software development—the model for the distribution of the level of effort becomes a log-normal distribution.

The beauty of this system lies in the ability to use the data later for modeling project plans. Even a simple spreadsheet—Excel,® for example—evolves into a model for the progress of a plan. The development team can model each target using the historical probability distribution functions for that particular task and then feeding those times into the next dependent task. Indeed, this method of analyzing schedules and budgets mimics the Program Evaluation and Review Technique (PERT) method created almost 50 years ago for the U.S. Navy's Nautilus submarine program. The Navy PERT planners used estimated values for pessimistic, nominal, and optimistic completion times and weighted them based on professional experience. The statistical approach—project simulation—we recommend relies on historical data and represents a better empirical model than educated guesses.

The following tools provide the power to build probabilistic models:

- Mathcad®
- Mathematica®
- Maple™
- S-Plus®
- R
- MATLAB®
- Scilab

Any good quantitative tool allows the mathematically oriented program manager to build a simulation of the project. In turn, the simulation provides important data for decision support; for example, the variance in expected project completion dates.

1.1.5.7 Example of a Phased Project

As we indicated, per APQP, automotive development projects always decompose into stages. In addition, the consensus management areas work within the automotive framework. A phase or stage gate provides the terminus for each stage. Each of the project segments illustrated below clarify and refine the project output. These segments reduce the risk to the organization by providing a map for the project team, a clear set of deliverable products, and criteria for project success.

Figure 1.5 provides one illustration for the gate review order. The review points may not happen as illustrated and sometimes evolve into a combined event; for example, after the product and process development phases when linked as shown.

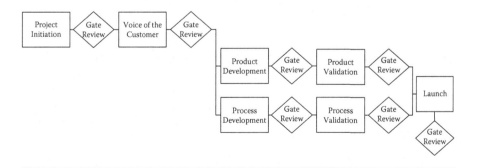

Figure 1.5   Phased approach.

We can structure the project phases in different ways. The phases and their purposes are the result of organization choices, such as

- Priority assessments,
- Product delivery processes,
- Cost control philosophies,
- Risk philosophies,
- Needs of other parts of the organization, or
- Needs of end customers.

For the sake of consistency, we present the project process described by AIAG, which has the following steps:

- Voice of the customer,
- Product development,
- Process development,
- Product verification,
- Process verification,
- Start of production, and
- Launch.

Later chapters handle each of these phases. We discuss the objectives and typical actions to achieve those objectives within each chapter.

It is important to note, while a single project manager may have to handle each of these phases, it is equally possible that there are project managers for each phase in the project. For example, for the early project phases, having a project manager who is skilled in the art of generating a number of possible solutions to the design could be beneficial. Through the production phase, it may be worthwhile to have the project run by a person skilled in the production processes. The following sections illustrate an automotive model for phased approaches to delivering a project. However, any industry can benefit from this approach—including the service industries.

Concept study   This phase provides multiple ideas for achieving market and organizational targets. This phase starts the creative phases—producing various concepts with the potential for achieving organizational goals and for identifying market constraints. We see examples of gate-passing questions for this phase in what follows:

- What are the markets in which the proposed product will complete?
- What is the market size for the proposed product?
- Have we identified a particular market segment to achieve a focused effort?
- Are there competitors in these markets? who are they? what can be learn from them?

- Does one of the proposed concepts better meet these product objectives and market goals?
- Does the estimated development and piece cost fit targets?
- Does one of the concepts present lower risks?

Clearly, the team can tailor these questions for embedded development and services.

Detailed development   During this phase, the engineering members of the team refine and document their design solution. Detailed development applies to both product and process design, as a process of progressive refinement occurs to produce the desired result. We reflect on the following examples of gate-passing questions for this phase:

- Do the specifications for the product fulfill product targets?
- Do the estimates for development meet available resources?
- Is the development time consistent with the product availability needs?
- Does the product meet organizational financial requirements?

Tailoring for embedded development and services is again very simple.

Final development   During this phase, the team refines and documents the selected design solution. Final development applies to both product and process design, since the process of progressive refinement moves to closure and final product release. Examples of gate-passing questions for this phase are the following:

- Does the developed product meet design specifications?
- Does the product meet financial requirements?
- Does the estimated quality and reliability level meet the market requirement?

Manufacturing development   In this phase, the manufacturing function uses the qualified design to exercise production line processes to verify they can deliver the product at required quality levels and production volumes. Much of the manufacturing process development can and should happen during the prototype, sample, and pilot units. Waiting until the qualified design to start the manufacturing process does not work. To qualify a design, the team will make a product with techniques and processes that represent production. Manufacturing or process development cannot be done in parallel; however, the team will want to minimize the variation.

In some enterprises, these steps include pilot runs, runs-at-rate (to determine cycle times), and start of production. In a service company, the development team exercises the service in a pilot market, executing the service in more than one market, and final launching of the service. Examples of gate-passing questions for this phase are the following:

- Does the production line manufacture products that meet specifications?
- Can the line produce the desired volumes?
- Can the line build the product that meets first run yield?

The team can tailor these questions also.

1.1.6 Project Management

Project management collects activities and tasks to achieve a set of goals. These tasks go beyond the execution of design responsibilities to include plans, schedules, and maintenance. In the most general terms, project management behaves as a control system that includes monitoring and corrective actions designed to minimize the risk of failure while driving toward the goals.

Another view of project management asserts a discipline for defining and achieving targets, while ensuring the effective and efficient use of resources. These resources include skills, information, time, money, locations, tools, and people. Project development lies within the domain of a project manager who often has little responsibility for the identified activities that produce the result. The project manager makes headway by facilitating interaction of the assigned project resources, removing road blocks, and promoting the understanding of project goals by team members. These activities occur to reduce the risk of failure while achieving the targets for the project scope and for quality within schedule constraints. When the project manager also carries technical responsibilities or must produce deliveries for the project, risk increases. In short, we suggest the job of the project manager lies in managing the project rather than acting as a technical team member.

Frequently, the team will see significant effects from late delivery, particularly when the project supports a regulatory change. Delivering an engine that meets a new U.S. Environmental Protection Agency (EPA) requirement by the government-imposed date remains important to automotive enterprises, because they will have no revenue and no profit if legal restrictions forbid the sale of their products.

The project management triangle, shown in Figure 1.6, retains relevance through the life of the development: the project scope, schedule, quality, and cost contain the product development and delivery aspects. Should the volume of the product change (for example, new features), the schedule, quality, and cost boundaries will change.

The consensus approach identifies a management area, called *planning*, as illustrated in Figure 1.7. This area defines those project aspects that

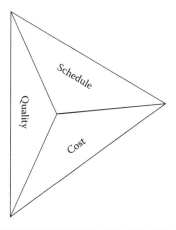

Figure 1.6 Project scope triangle.

support managing and achieving the time-related goals of the project. Project schedule estimation lies within the domain of planning. The following figure shows the planning interaction with the project. The project manager and team must have schedule integrity. Figure 1.8 defines the process interactions that ensure the quality of the product. While this figure does not capture every input, it captures the major project process deliverables and interactions, including documentation to identify and ease the achievement of quality targets. The quality assurance plan documents the activities that maintain the quality of the project and subsequent product (applies to both services and manufacturing).

Figure 1.9 illustrates the management area for cost control. The actions executed in this area determine the resources required and the estimating and budgeting activities. The work breakdown structure (WBS) represents a key input element to this process since it defines required activities for each cost center. The cost control area uses earned value management techniques to monitor the state of the project toward the plan and budget.

The definition of the project boundary represents the project scope. The project scope defines the project objectives *and* the deliverable products—it is developed as a result of the initial requirements analysis and functional allocation. Figure 1.10 illustrates how we define processes for identifying and controlling the project. The project charter, scope statement, scope containment, and the WBS belong to the scope definition as key outputs.

Project scope, objective, and deliverables must be logical. The project manager cannot assign a task to a part of the team that does not have the capability to execute the task. For example, the team should see no logic in assigning a task that requires software development to a team member when the contract does not allow that person access to software. The project

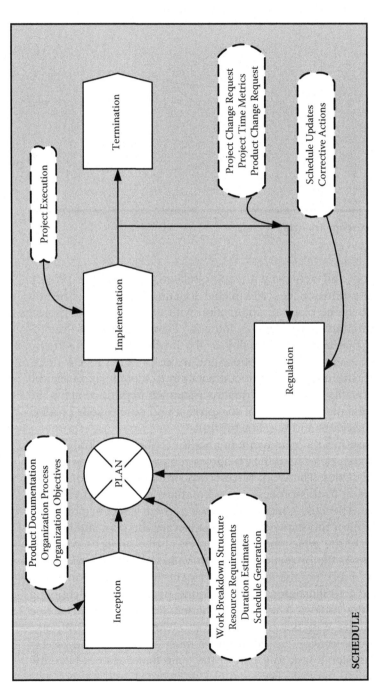

Figure 1.7 Planning/scheduling.

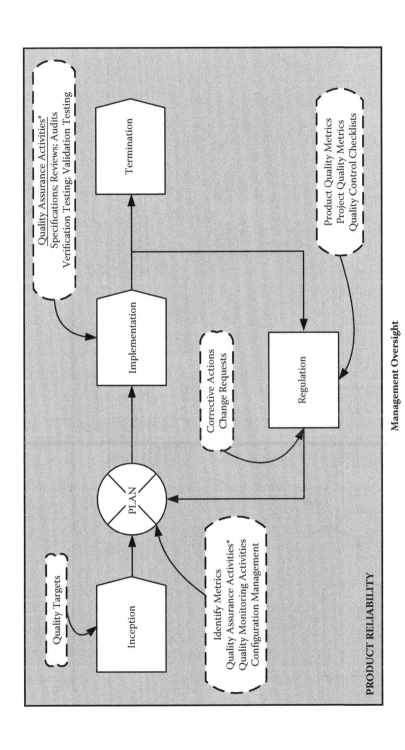

Figure 1.8 Quality and product integrity.

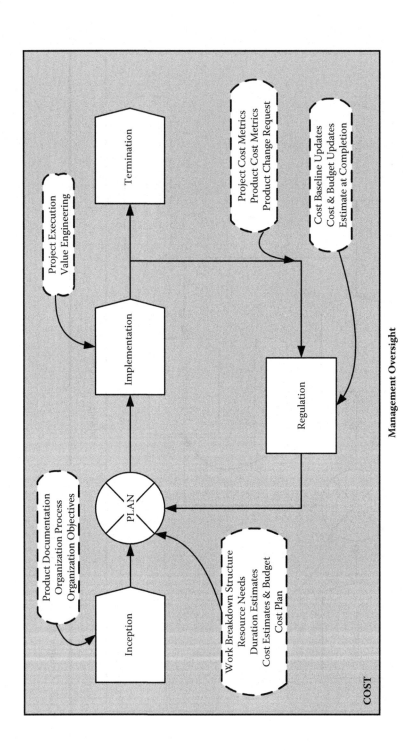

Figure 1.9 Cost control.

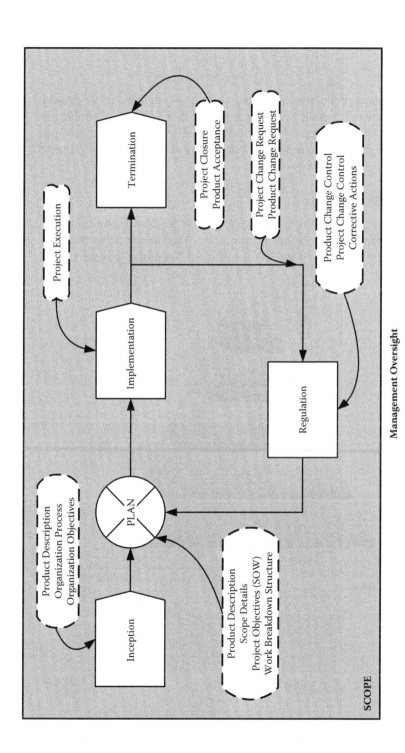

Figure 1.10   Scope control.

manager and the team should access and reaccess realities when it comes to scope, objectives, and deliverables.

1.1.6.1   People

Resource allocation becomes a critical aspect of delivering a project because projects work within limited resources; that is, they are finite capacity processes and the team should understand and model them this way. Limitations range from monetary constraints to access to specialized resources (exotic skill sets).

1.1.6.2   Limited Resource

In our experience, the resource issue grows into the most common stumbling block to achieving goals (see Figure 1.11). This issue seems to occur in situations where the enterprise has insufficient resources (capacity) and

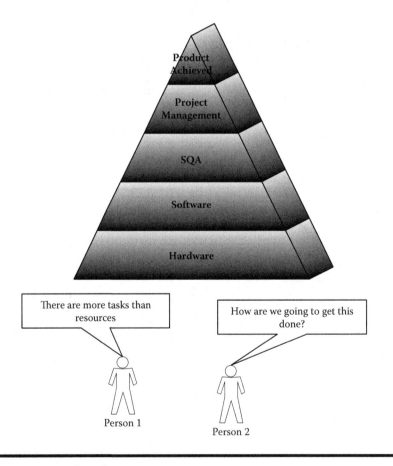

Figure 1.11   People and resources.

resources (people) must time slice their effort in order to provide deliverable products.

Additionally, pressure to make efficient use of resources and reduce redundancy creates an environment from which it becomes difficult to recover when key people go on vacation or leave for other opportunities. Mature organizations try to manage the loss of knowledge by establishing succession plans that spell out the availability of skills and provide for stepping in by other individuals.

Resource provision   This section of the standard requires that the compliant organization identify their resource requirements explicitly. Additional considerations relate to the allocation of resources once they have been identified. Naïve identification of a resource does not make that resource available for tasking.

Human resource constraints   The program manager coordinates with human resources the moment he chooses a team for the project. At the onset of the project, the team behaves as a barely-unified collective than one might call a "team." The bonding required to create an effective team takes time and, frequently, exertion. Additionally, common interests and shared experiences can hold teams together. Figure 1.12 provides an approach to managing human resource constraints.

Infrastructure   The resource infrastructure consists of the hardware, knowledge, and services that support the project members. The lack of proper tools—even something as ephemeral as knowledge—hamstrings project progress and increases quality, schedule, and cost risk.

Work environment   The work environment must be conducive to producing a quality product. Everything from appropriate tools to human factors issues is under this heading. Tools can involve location and hardware/software and services needed to accomplish tasks. If development work occurs outside the United States, the developers may find it difficult to purchase certain kinds of regulated software.

1.1.7   Project Manager's Role

Project managers share the same responsibilities whether they work in automotive or nonautomotive industries. The primary difference in automotive project management lies in the plethora of quality documents and tasks demanded by the automotive marketplace and the ISO/TS 16949 quality standard. Figure 1.13 illustrates the project manager's role within the organization. In the organization below, the management layer of the respective functions coordinates the project. The project manager does not bear the bulk of the burden to deliver the project.

A matrix models the organization as shown in Figure 1.14. In the matrix organization, the project manager has people who report to him from various parts of functional departments. These people report to both the project

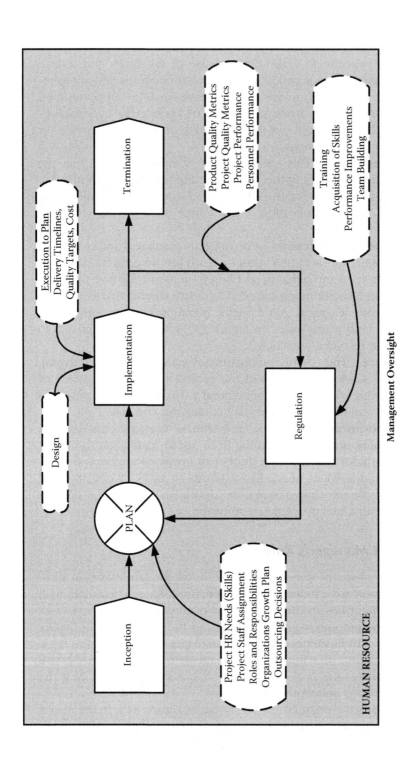

Figure 1.12 Managing human resource constraints.

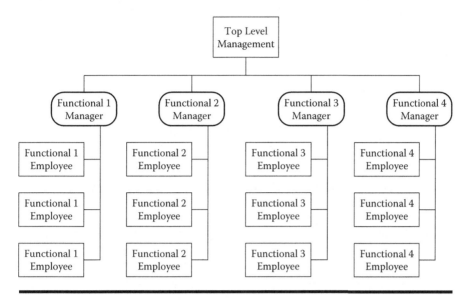

Figure 1.13   Functional organization.

manager and the functional manager. This dichotomy can negatively affect the employee and requires much communication between the project manager and the functional managers. In this organization model, the project manager has more responsibility and a higher set of expectations than in models that favor functional management.

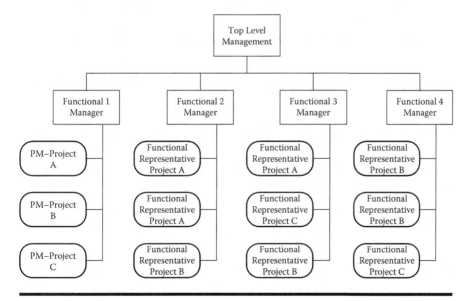

Figure 1.14   Matrix organization.

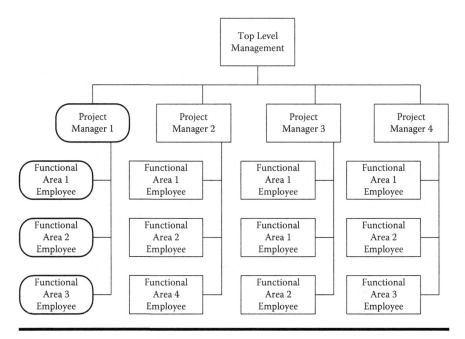

Figure 1.15 Project organization.

In the organization type in Figure 1.15, the project manager coordinates the project with the functional areas represented by those working on the project. The project manager has the responsibility of reporting the status of the project to management and has the majority of the control over the human resources of the project.

It is important to know where your organization and any supplying organization fit on this continuum (see Figure 1.16). These organizational structures identify and, to some extent, dictate authorities and responsibilities. Solving an issue via the project manager of a functional organization may not be the best approach for timely resolution of a project concern.

1.1.7.1 Organization and Integration

A good project manager establishes communication with all stakeholders and provides an update mechanism to keep them involved and aware of

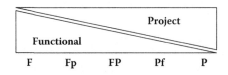

Figure 1.16 Project organization.

project progress—enhancing meaningful communication improves organization and integration of development. In January 1996 the Gartner Group, in its paper Project Management Skills: Avoiding Management by Crisis, identified insufficient involvement of stakeholders and infrequent communication with sponsors as leading causes of project failure.[2]

However, this does not mean that all communications or even the bulk should be focused upon the stakeholders. The project manager must be able to communicate throughout the project. *The Little Black Book of Project Management* [Thomsett 2002], by Michael C. Thomsett, identifies five areas of communications breakdown:[3]

1. Team member to team member
2. Project manager to team member
3. Project manager to outside department manager
4. Manager to outside resource
5. Manager to executive

Communication challenges are not exclusive to automotive development. However, automotive development often requires input or project participation on a global basis, where the time zone differences alone provide added stress. The role of communications in project management finds expression in the *36 Hour Course Project Management* [Cooke and Tate 2005], where you cannot automate the clarification of assumptions, integration of inputs, and the iterative process of analysis.[4]

Kim Heldman in her book *Project Managers' Spotlight on Risk Management* [Heldman 2005] says, "communicating is the most important responsibility you have as a project manager. Ninety percent of your time is spent in this activity. I can think of no other activity that has a greater impact upon project success."[5]

One model of communication [Shannon, 1948] pictures information flow traveling over a *channel*. An equation defines the number of communication channels (the number of interconnections) and the result relies on the number of people required to interact.

$$\textit{Number of communication channels} = \frac{N * (N - 1)}{2}$$

The communications plan articulates the communications requirements for the project. A communications plan helps to facilitate communications among the required parties, particularly when an issue needs escalation to a higher level of management. For small projects—those projects with few stakeholders—this formality dissipates. The project team should consider how information distributes among participants. The team specifies activities for creating and reporting decision documents and consults with customer and stakeholder for their responses. The project manager chooses procedures for creating, storing, accessing, and presenting information. These acts define what information the team needs and the format required by the project participants, including stakeholders.

The typical contents of a communications plan comprise the following:

- **Information distribution**   A description of the communications distribution method—often a chart detailing the communications responsibilities. ISO/TS 16949 requires the inclusion of customer-specific documents.
- **Information description**   A description of the type of information for distribution, format content, and level of detail.
- **Information schedule**   Method for assessing and rates of delivery of information.
- **Progress reporting**   A description or reference of the process for collecting and reporting project status or progress reports.
- **Communications plan revision**   Method for revising, updating, and otherwise refining the communications management plans as the project progresses.
- **Administrative closure**   Generating, gathering, and distributing information to formalize a phase or project completion. Administrative closure consists of documenting project results to formalize acceptance of the proposed or completed product by the sponsor or customer. It includes collecting project records, ensuring that they reflect the final specifications, analyzing project success, recording effectiveness and lessons learned, and archiving such information for future use.

The team's organization and integration activities identify project requirements for each process of the project. Figure 1.17 illustrates the actions needed to fulfill the control of communications during a project, which leads to enhanced organization and integration of the team. Communications link the project, the customer, and the supplying organizations into a social network. It provides feedback (in the control process) to the organizations regarding the status of the project.

1.1.7.2   Customer Focus

The project manager focuses on the client or customer, particularly in customer-focused product lines. Getting to know the customer, communicating often and clearly, and cultivating a personal relationship with customer representatives functions as key methods for achieving goals due to increased opportunities for influence and persuasion. From the customer's

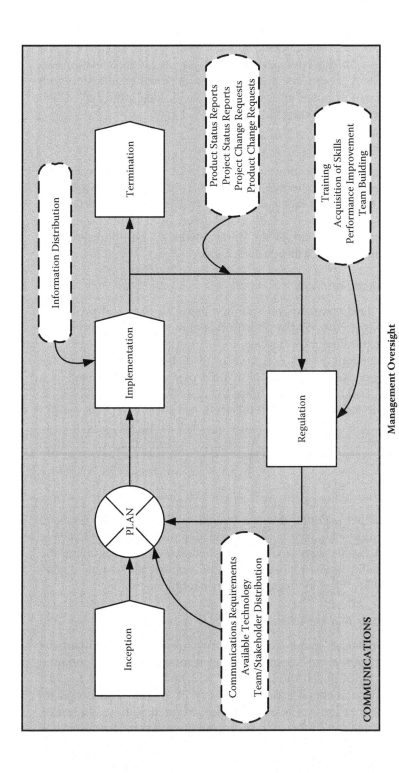

Figure 1.17 Using communications to integrate development.

point of view, the project manager embodies the project and serves as the principal point of contact. One ultimate objective of strong personal relationships lies with the development of an organizational relationship with the customer. A project manager without a personal relationship with the customer metamorphoses into a note taker with little influence, which reflects poorly on the organization.

1.1.7.3   Brainstorming

Brainstorming and mindstorming are, respectively, group and individual techniques for generating potential solutions for a designated problem or goal. Typically, the session facilitator solicits criticism-free suggestions in order to increase the quantity of ideas followed by a reorganization of ideas, most often in the form of an affinity diagram. The brainstorming team follows diagramming with idea selection.

1.1.7.4   Technical Solutions

The project manager possesses no role in solving all technical problems. The project manager ensures that a technical solution can be found and, if not, manages stakeholder expectations.

Often, management passes on projects to technical people with the belief that a technically competent individual performs equally well as a project manager. Technical prowess helps but does not guarantee success as a project manager. It seems that often this technical ability keeps the project manager in the realm of the technical, forcing or dictating solutions instead of performing project management work and evoking the best solution from technically responsible individuals.

1.1.7.5   Quality

The project manager accounts for the overall quality of the project and integrates this work with various functional departments. The challenge lies in balancing quality against cost and schedule without degrading the product. The quality scenario becomes more complicated when quality over time (reliability) derives from a customer requirement.

1.1.7.6   Facilitate Team Creation

While a project manager may receive a staff from management, that does not mean that a team exists. Teams develop through a body of common experience and shared goals, and management cannot force a team to exist. The project manager enhances the probability of success and decreases project risk if he can grow a team from the assigned staff. Additionally, the project manager emphasizes the responsibility the team has regarding delivery of the final product.

Key characteristics of an effective team are

1. Strong team identity: having a team name and the rest of the organization knows of the team,
2. Uniqueness: feeling like "we are extraordinary,"
3. Commitment: feeling ownership in a project—that is, buy-in,
4. Competency: acknowledging team competencies, and
5. Fun: creating a fun-loving environment

Bad teams may also happen. Neither type of team occurs randomly, but rather as the result of organization dynamics and managerial involvement. Some key attributes or conditions that spawn the creation of a poor team are

1. Lack of trust among team members,
2. Unfocused time spent on multiple projects,
3. Incompetence and lack of appreciation for the capabilities of other individuals,
4. Arbitrary deadlines,
5. Misdirected communications,
6. Lack of respect for other teams or parts of the company, and
7. Teams within teams (factions).

1.1.7.7 Conflict Resolution

Project managers may have to resolve intrateam conflicts. Not all conflict becomes counterproductive and not all conflict requires mediation from the project manager. Guffey [Guffey 2003] identifies two types of conflict: *cognitive conflict* and *affective conflict*.[6] Cognitive conflict focuses on issues and on developing good creative solutions to problems as the dialectical interplay yields higher order reconciliations of ideas. Affective conflict focuses on feelings and personalities. The project manager understands when the conflict produces a negative effect on the team and facilitates resolution by

1. Avoiding blame or scapegoating,
2. Clarifying and defining the issues,
3. Listening intently to each party,
4. Stating points-of-view clearly,
5. Working on points of mutual agreement,
6. Brainstorming or mindstorming alternate solutions,
7. Agreeing on a potential solution,
8. Documenting the agreement (making a contract), and
9. Auditing the solution for efficacy.

1.1.8 Product Development Team Meetings

A good project manager includes more than the design engineering staff in the development of the project. With manufacturing input from the start, the team should find it possible to design the product in a way that takes advantage of the strengths and existing methods and tools of the manufacturing facility. Cross-functional team meetings lead to actions within the organization to coordinate delivery of project results to production. These actions occur in the final phases of the design when the design becomes solid and ready for production tooling.

Circumstances generally require *some* team involvement at each stage of the project or program for prototypes, samples, or pilot runs. The project manager balances design with manufacturing. During the commencement of the prototype build, manufacturing may review and provide feedback for consideration, but as the project advances manufacturing receives more weight. At the pilot stage, manufacturing should correspond with design. Few programs today have the luxury of making design changes just for manufacturing. The manufacturing process may trod one step behind design, a situation not uncommon during concurrent engineering. Note that these situations can also affect service products when implementation and planning don't synchronize.

Representatives from different functions of the organization make up a diverse team. Typically, the team members include the following:

1. Project manager from supplier (when applicable)
2. Internal project manager
3. Supplier quality assurance (SQA)
4. Shipping and receiving (packaging)
5. Production line representative
6. Design
7. Documentation
8. Manufacturing

The project manager does not have to require all participants to attend all meetings; rather, each area of expertise participates in its action items. The design of these meetings coordinates the controlled introduction of the product to the production line efficiently while minimizing disruption. The meeting should not evolve into a forum for solving problems. Occasionally, when a problem requires interactive collaboration, the meeting emerges as a tool for team problem-solving. Like most meetings, team members should collect the information *before* the meeting, distribute it during the meeting, decide on it, and follow with action.

1.1.9 Program Management

We define a program as a collection of projects executed concurrently. These projects require coordination and simultaneous delivery to meet overall objectives. A master schedule helps synchronize these subproject schedules. The master schedule provides structure for the deliverables for the subordinate projects. Vehicles undergoing major changes or newly created vehicles require the coordination of components to one delivery schedule, a specified cost, and program and project quality requirements. The same approach would apply to any program designed to deliver a new service.

1.1.10 Embedded Development

This book adds an emphasis on development of embedded or software projects. With an embedded project, the development team has a significant portion of the development work in software located on the product itself (some form of nonvolatile memory). An embedded development project consists of software development for a microcontroller or microprocessor-based hardware. The features of the product developed reside in the software application. The software can consist of an operating system for a complex project along with an application code. Product-specific features reside in the application code. The software and the hardware differentiate poorly (hence the frequently used neologism "firmware"). Firmware weds the hardware and software because, on its own, the hardware has no function. The engineers work the hardware and the software development simultaneously (see Figure 1.18) in order to achieve a form of inorganic symbiosis; that is, the final product results from a dialectic exchange, a back and forth adaptation of each component to the other.

Software possesses its own problems; for example, the increase in complexity of the software product as the number of paths and variable values increases with added features. Version control, testing, and good configuration management (see Figure 1.19) help to verify, validate, and control embedded software.

1.2 Product Integrity and Reliability

With the idea of *product integrity*, we refer to a more general concept of integrity: a oneness, an entity with no discontinuities, a form of honesty. Reliability refers to quality expressed over some unit of time and typically has overtones of durability and robustness.

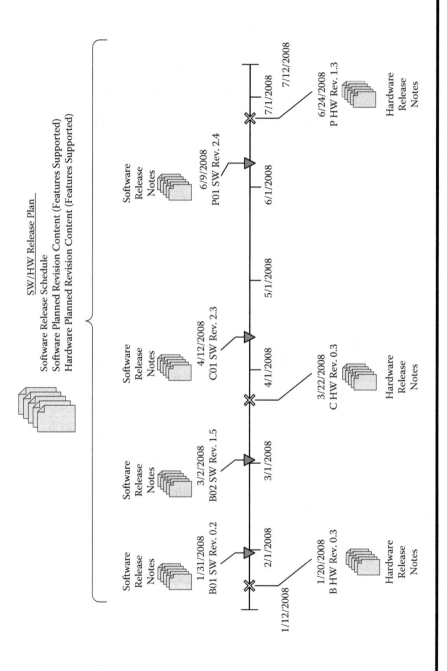

Figure 1.18 Configuration management.

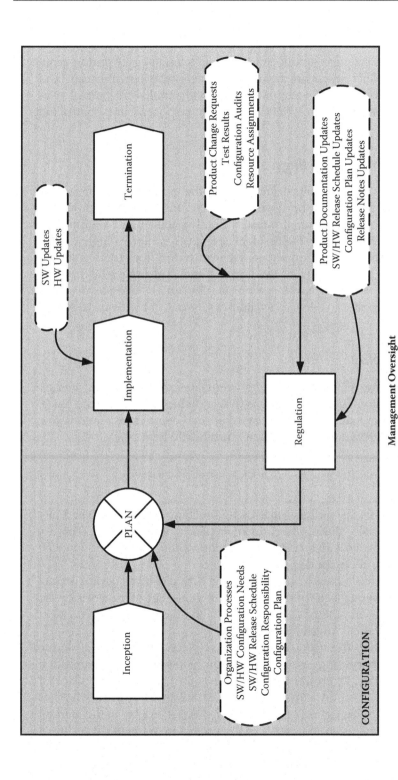

Figure 1.19 Configuration management.

1.2.1 Product Integrity

Stage targets can promote product integrity and team integrity during every phase. The product and team retain their integrity when the team and management review the development work with ruthless rectitude, stripping out unnecessary features, defining realistic next steps, and controlling the scope of the work.

1.2.2 System-Level Testing

System-level testing normally occurs during the end game portion of a project or program. Then, the developers have enough production-level material to perform a meaningful test with all modules and submodules exercised. This event acts as the first real attempt to achieve a high level of verisimilitude. Individual components receive testing from tools that simulate the target system or from existing systems converging to the final product. Testing should occur as early as feasible and as often as possible. This approach applies to services as much as it does to hardware and software products.

1.2.3 Reliability

As mentioned above, reliability refers to quality over time. The product must meet requirements/specifications over a defined time. In some cases, parts of a product may degrade; for example, end users expect automotive tires to degrade with time—in effect, a consumable item.

1.3 Cost

Cost estimation begins with the purchasing function, sometimes called "acquisition," especially during the formative phase of the project. The accounting function provides cost controls during the course of a project. Cost targets inform design decisions.

The delivered to estimates graphic, Figure 1.20, illustrates typical project expenditures and payback for those expenditures—part of the decision for accepting or rejecting the project. In short, when the results of the project add to the profitability of the organization.

Underestimating the needs for the project places the project at undue risk by potentially causing resource starvation. The estimating organization may lose credibility. Overestimating the project cost alienates the customer due to poor return on investment. Figure 1.21 illustrates a project that has a cost overrun. In this example, the situation has put a delay in the amount of time for the organization to recover the project investment. In reality, if the cost overrun becomes too high, the project may have no payback at all.

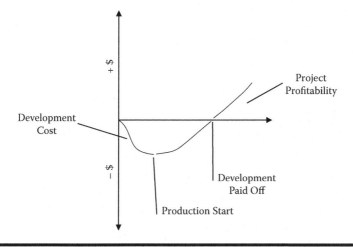

Figure 1.20   Delivered to estimates.

Figure 1.22 illustrates a schedule overrun. Late delivery of the product delays payback for the organization, a condition frequently referred to as "opportunity cost." In the example below, the team would see no associated cost overrun. In reality, if the situation delays the schedule, a high probability exists for a cost overrun. In this instance, the organization launches the product late, with some potential opportunity cost.

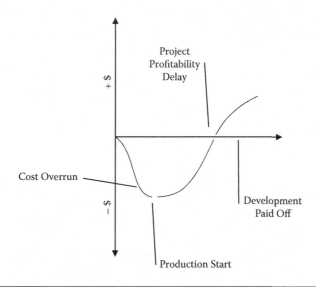

Figure 1.21   Cost overrun.

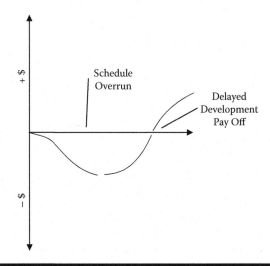

Figure 1.22   Schedule overrun.

In Figure 1.23, the product birthed on time and within development budget. However, the product cost ended up being higher than anticipated and the profit margin on each unit became lower than anticipated. This undesirable situation prolongs the time it takes for the company to pay off the development work and begin profiting from the project.

In order to know how much profit returns to the company when quoting a project, the team must understand the level of effort it takes to deliver the project and the product cost. If the enterprise makes money conducting

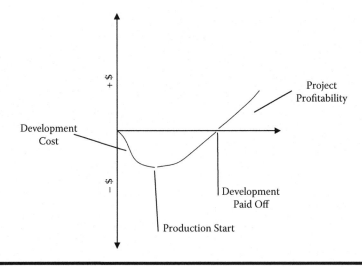

Figure 1.23   Component cost.

the project, but not in selling the resulting product from the project, then product costs will not meet target values well enough to generate the anticipated margin. How does the value of the part change? In manufacturing enterprises, the project manager will invoice during-project parts at a prototype rate much higher than the actual material cost, while invoicing the production parts at a price set related to a corporate hurdle (e.g., internal rate of return (IRR), net present value (NPV), payback, or any combination of these). A fixed cost contract means the supplier owns the risk of producing the project results within the defined cost, since no opportunity for recovering the cost exists. Cost plus firm fixed-fee contracts, on the other hand, allow the supplier to recover development costs.

1. Development cost
   a. The make or buy decision
   b. Value engineering
   c. Material composition
   d. Development processes
   e. Product specifications (can also apply to manufacturing)
      i. Product standardization
      ii. Tooling cost
      iii. Material sourcing
      iv. Material substitution and obsolescence
   f. Verification requirements
2. Sourcing
   a. Travel
   b. Human resources
3. Manufacturing cost
   a. Manufacturing equipment
   b. Manufacturing location and facility
   c. Manufacturing material
   d. Manufacturing process
   e. Manufacturing verification requirements
4. Maintenance cost
   a. Extensibility/obsolescence
   b. Product life cycle
   c. Manufacturing material
   d. Manufacturing process

1.4 Structure of Sections

This book lays out the program sequence analogously to the sequence specified for APQP. The chapters are

1. Projects and Project Managers
2. Technical Tools

3. Concept
4. Product Development
5. Process Development
6. Validation of Product and Process
7. Release to Production
8. Failure Reporting Analysis, and Corrective Action System (Phase)
9. Product Support
10. Program Management
11. Final Thoughts

Chapter Notes

[1]Automotive Industry Action Group, *Advanced Product Quality Planning and Control Plan* (APQP), (Southfield, Michigan, AIAG 1995) p. 5.

[2]Project Management Skills: Avoiding Management by Crisis, The Gartner Group, http://www.projectsmart.co.uk/stakeholder-management.html, January 1996 (accessed February 24, 2008).

[3]Michael C. Thomsett, *The Little Black Book of Project Management* E2, (New York, AMACOM 2002) p. 47.

[4]Helen S. Cooke and Karen Tate, 36 *Hour Course Project Management*, (New York, McGraw-Hill 2005) p. 123.

[5]Kim Heldman, *Project Managers Spotlight on Risk Management*, (San Francisco, Harbor Light Press 2005) p. 20.

[6]Mary Ellen Guffey, *Business Communications: Process and Product* E4, (United States, Thomson South-Western 2003) p. 47.

Chapter 2

Technical Tools

2.1 Delivery

2.1.1 Axioms of Program Management

We start off this chapter with a tongue-in-cheek collection of "axioms" that sum up some of the more critical realizations we had during numerous projects.

1. All time lines belong to a directed graph
    a. The network diagram has more importance than the Gantt chart because it more adequately represents the relations of the tasks and deliverables.
    b. Microsoft Project® does a poor job of supporting the network diagram.
        i. Fix this situation with PERT Chart Pro® from CriticalTools.com, a Project® plug-in.
2. To calculate a critical path correctly, we must have
    a. One entrance point,
    b. One exit point,
    c. All other tasks connected to at least one other task at the beginning and end of the task,
    d. Program managers who do not use the term "critical path" unless they understand what it means,
    e. Use Microsoft Project so that it does not provide an illusory critical path-BEWARE!
3. Baseline all time lines once the plan has approval.
    a. Alter no plan without exposure at the executive level.
    b. All functional managers must deliver a complete plan no later than the planning gate of the project.

      c. Measure plans according to the standard metrics as defined in chapter 15 of Kerzner's *Project Management*, 8th or 9th edition [Kerzner 2001].

         i. Use earned value analysis—supported by Microsoft Project and Primavera Project Planner® (p3).

        ii. If payroll dollars become a touchy topic (e.g., salaries), then use hours as a substitute.

      d. Failure to meet the plan equals an annual evaluation issue.

4. The project manager should create plans at as fine a granularity as possible so that the completion of tasks becomes a binary choice and the percentage completion indicator of the Microsoft Project software actually means something.

5. Program managers should manage deliverables not tasks.

      a. Functional managers hold responsibility for task completion.

      b. Delivery either exists or not (binary).

6. Hard-schedule all gates when the team agrees to take on the business.

7. All of the consensus management areas are the responsibility of the project manager, not just arbitrarily laid out in the schedule.

8. Any launch process serves us better than no launch process.

9. Build slack into the time line from the start of the program and manage it with great care.

2.1.2 Supplier Selection

Whether the suppliers are outsourced services or actual manufacturing suppliers, they always remain significant to the project because they participate just as much in the result as any other resource. In some cases, the corporate customer may dictate the choice of suppliers, which can lead to major problems when those suppliers become unreliable.

The bases for choosing a supplier can vary, as in the following:

1. The supplier presents a service or part concept and customer selects them.

2. Internal engineering or process design generates the concept and out-sources to the supplier with appropriate supporting information.

3. The enterprise selects the supplier due to an ongoing relationship.

The next section illustrates various methods of evaluating a supplier. We use a combination of in-house methods and standards. However, while the acquisition function selects the supplier, the project manager should know and assess the risk involved with each part or service and with each supplier.

2.1.2.1   Supplier Evaluation

Selection of a supplier relies on numerous factors—many of the evaluation criteria depend on the economic performance and the stability of the supplier. Some companies have effective evaluation methods for the economics of the organizations. Other organizations also have effective engineering evaluation criteria. However, often a gap exists in the evaluation standards, particularly for embedded software development. Key development tool requirements may not appear in these supplier evaluations.

In the case of services, obvious presentation of requirements in the form of mechanical drawings makes no sense. A service company may need to create a specification or a statement of work to provide enough information for an outsourced service to provide a quote.

The evaluation grades the supplier's capabilities. For each category, there may be multiple choices to quantify the supplier's capability with respect to project requirements. The evaluation team will associate a score with each of these possibilities, particularly in the case of government contracts. The sum of these scores represents the supplier's capabilities. The supplier evaluation does not select the supplier; rather, the scores developed during the acquisition process provide an ordinal list of supplier capabilities. In the automotive industry, a group consisting of representatives from the Supplier Quality Assurance (SQA) function, technical expertise from the design staff, and the acquisition function (for example, the purchasing department) performs the supplier evaluation. Team members should participate in this evaluation in order to provide the project manager with a preliminary understanding of the strengths and weaknesses of the supplier.

In the case of software acquisition, internal methods of selecting suppliers often do not evaluate the supplier in key software practices. Instead, the review or critique relates more to the supplier's financial and production constraints. The choice of software supplier based solely on financial data can be myopic and reflective of insufficient technical involvement in the acquisition activity.

The following list provides some of the factors for consideration in the supplier evaluation:

- Company ownership
- Affiliated organizations or parent organizations
- Facilities (global or regional)
- Sales turnover
- Net income
- Management expertise
    - Customer satisfaction
    - Risk philosophy

- Production material
  - Material management
  - Logistical systems
- Organizational structure
- Customers (most volume)
- Organizational awards
- Quality system
  - Quality philosophy
  - Quality planning
  - Quality assurance methods
  - Problem solving methods
- Research and development expenditures
- Existing product Parts Per Million (PPM) figures
- Existing product warranty statistics
- Historical product and project delivery information
  - On-time delivery
  - Cost of project
- Tools
  - Computer aided drafting (CAD)
  - Computer aided manufacturing (CAM)
  - Simulation and emulation tools
  - Verification tools
- EDI capabilities
- Supplier reliability
- Product development
  - Development personnel (skills)
  - Product development processes
  - Development tools and systems
  - Prototype availability
  - Design change handling
- Project management
  - Organization
  - Project processes
  - Human resources
  - Project change management
  - Subcontractor performance management

2.1.2.2   Capability Maturity Model Integrated

Developers use the capability maturity model integrated (CMMI) method in evaluating software or engineering suppliers. The model purports to assess the *maturity* of various tasks within an organization—not only software—and applies a score. A complete auditing standard exists for this approach. However, little research supports this model as a significant evaluation tool

Table 2.1   CMMI Levels and Characteristics

Maturity Level	Level Name	Process Characteristic
1	Performed process	Chaotic
2	Managed process	Disciplined
3	Defined process	Repeatable
4	Quantitatively managed process	Controlled via statistics
5	Optimizing process	Continually improving

for assessing the software development of a supplier; that is, the project manager or team can assess maturity of the development process, but it cannot directly assess the software itself.

2.1.2.3   Software Process Improvement and Capability dEtermination

Software Process Improvement and Capability dEtermination (SPICE) has evolved into ISO/IEC 15504, a model similar to the CMMI. In fact, the SPICE effort in Europe probably influenced the older capability maturity model (CMM) to evolve into the CMMI.

2.1.3   Work Breakdown Structure

Initial scope containment actions identify those activities needed to ensure the scope of the project does not submerge within the processes of the project.

The work breakdown structure (WBS) takes the top-level deliverables of the project and functionally decomposes these items into a hierarchical representation of the final product. In U.S. Department of Defense (DoD) vernacular, the WBS provides *cost centers* for cost and schedule tracking of the project. The team should refer to the lowest element in the WBS as a *work package*. The decomposition of tasks needed to produce the project objectives allows for detailed estimations of project costs. Additionally, the team can match the work packages against available resources to provide a more complete assessment of the feasibility of the project. Decomposing cost centers to some *atomic* level, for example, where we have estimates between eight hours and eighty hours usually improves the accuracy of the forecast. What follows is a benefits list for any WBS when allied with a bill of resources:

1. Breaks the project down into the lowest components
2. Helps with the development of duration estimates
3. Aids development of resource assignments and responsibilities (identifies skills and skill acquisition needs)
4. Facilitates risk identification and mitigation

5. Identifies tasks supporting activities for the design
6. Identifies tasks for support plans such as configuration management, quality assurance, and verification and validation plans.

In order to perform effective estimation of the duration of a task, the project manager needs an in-depth understanding of both the requirements and the required actions. Therefore, the estimates should flow up from those resources that execute these tasks; that is, the team members and their managers provide their own estimates. The estimates may be measured by:

■ Time, in the form of hours or days,
■ Money, in the form of dollars,
■ Person-hours, a combination form.

Work package estimation occurs more quickly with the use of a complete WBS—these atoms simplify the process of estimation because they depict small, comprehensible actions. Sometimes, the team member (executor) does not estimate, but rather a technical expert or the project manager will estimate. Bringing in a "pinch hitter" adds little value to the derivation of good estimates—the person who will complete the task should perform the estimates! In so doing, the team increases the likelihood of commitment to that task; participation in the estimation process encourages ownership of the results, converting the players on the team from victims to participants. Note that we posit work package estimation as a dynamic process designed to produce meaningful results. Having the project manager dictate the desired schedules to the team while ignoring contributions from team members demotivates the project team.

We see an example WBS in Figure 2.1.

2.1.4 Resource Breakdown Structure

After the planning team creates the WBS, the project manager in concert with the team will identify and assign the resources needed to undertake the individual work packages, identifying skills and assigning responsibilities. A resource allocation matrix (sometimes called a "bill of resources"), Figure 2.2, can help to convey the areas of responsibilities to the project team.

It may be naïve to believe that people assigned to the project work solely on their project tasks. Personal efficiency and normal interactions consume part of each working day, implying full utilization as impossible even under ideal circumstances. If a person works half-time on a deliverable, one can assume it will take at least twice as long to complete that task. In this case, the team assumes little or no disruption in the transition from the other tasks, a possibly unrealistic option.

The project manager would be wise to document utilization assumptions. These assumptions allow for more accurate predictions and also give

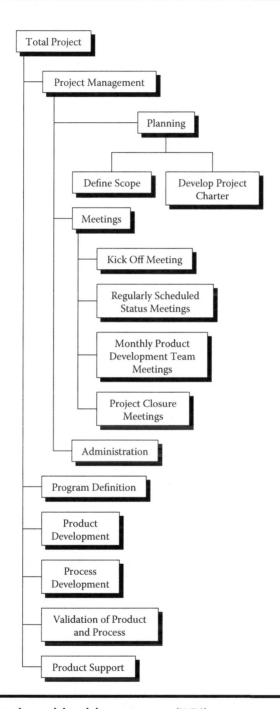

Figure 2.1   Part of a work breakdown structure (WBS).

	Component2 SW	Component2 HW	Project Manager	Component1 SW	Component1 HW	System Spec	System Verification	Tools	Manufacturing	Aftermarket	Training	Tech-Pubs
Design Doc												
Component2	A	In	A	In	In	In	In	In			In	In
Systems	In	In	In	In	A	In	A	In	In		In	In
Component1	In	In	In	A	A	A	In	In	In		In	In
Component2 Release												
Hardware	A	A	A	In	C	C	In	C	C		In	In
Software		C		In	C	C	In	C	C		In	In
Component1 Release												
Hardware	C			A	A	A		C	C		In	In
Software		C		In				C	C		In	In
Testing												
Component2	A	A		In	In	In	In	In				
Component1	In	In		A	A	A	In	In				
Systems	In	In	A	In	In	In	A	A	A			
Training												
Component2		A	A	A	A	A					R	R
Component1											R	R
Systems						A	A				R	R

Accountable — A
Participant — P
Review Required — R
Input Required — I
Sign Off Required — S
Informed — In
Consulted — C

Component2 SW — RD
Component2 HW — SR
Project Manager — AC
Component1 - SW — MB
Component1 - HW — MB
Systems Spec — WC
System Verification — MM
Tools — JG
Manufacturing — BW
Aftermarket — SR
Training — BB
Tech-Pubs — LO

Figure 2.2 Example of resource breakdown structure.

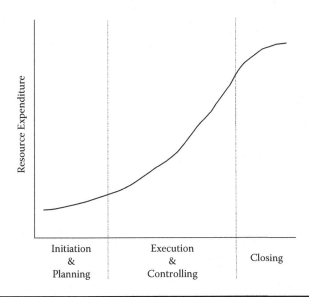

Figure 2.3 Example of accumulated Human Resource (HR) expense.

visibility to the actual workloads. Keep in mind that the cost and schedule assumptions represent a *model* of what the project manager desires. The project manager should be wary of cases where an individual with a penchant for overwork takes on all tasks and fails—the principal defect of infinite-loading models.

Our approach to the management of human resource contraints appears in Figure 2.3.

2.1.5 Project Estimating and Scheduling

When the project manager estimates a project with his team, he can usually estimate cost and schedule while setting target values for quality. Figure 2.3 shows an example of how the cost of a project accumulates.

2.1.5.1 Top-Down Estimating

Top-down estimating relies on historical project budgeting. The project manager can apply this method when the historical project attributes resemble the current project. If an instrument cluster development project always costs $2 million then this amount would be budgeted and distributed among project phases, distributed in the proportions suggested by past projects. Below we illustrate a possible budget distribution using this method.

Table 2.2 Budget by Phase (Top-Down Estimating)

Project Phase	Percent of Budget	Dollar
Voice of customer	5%	$100,000
Product development	30%	$600,000
Process development	30%	$600,000
Product and process validation	35%	$700,000
Total	100%	$2,000,000

2.1.5.2   Bottom-Up Estimating

Bottom up estimating rolls up the WBS task durations. Once the project manager and team assign a duration and cost to each task, the project manager compiles this information into a schedule and a budget. Individual team members participate in the bottom up approach, while higher-level managers and the project manager have editorial responsibility over the estimates, providing a filter against wild inaccuracies and simple mistakes.

2.1.5.3   Phased Estimating

As each phase terminates, the project manager and team revise or recalculate estimates with additional information from the previous phase. Each subsequent phase increases the details for the next and improves the estimates. In short, actual events consume the forecast (schedule and cost).

We describe project management as a process of progressive elaboration. In the early phases of a project, the entire team moves into the unknown. They may have nebulous scoping details. As events consume the forecast, the project manager replaces vague estimation with real data and the remaining forecast improves in quality.

If upstream management interferes with the project by dictating a compressed schedule or a reduced budget, the likelihood of a successful project diminishes. Unrealistic due dates degrade the quality of the schedule and unrealistic budgets degrade the value of project costing. Higher-level interference can destroy the sense of ownership in a team by shrinking the perception of participation and demeaning the contribution of team members.

Additionally, crashing (or reducing) the schedule generally fails to account for the effect of random variation on the project plan. In retaliation or expectation, some project managers react by padding their estimate; that is, inserting safety lead time to increase the likelihood of task completion. Unfortunately, padding produces a distortion in the estimates of both time and cost. An even worse situation occurs when the upstream managers begin to assume the project managers padded the budgets and routinely call for schedule and budget attenuation.

2.1.5.4 Project Evaluation and Review Technique

While some elements of a project may recur from project to project, such as a well-defined software release process, many elements occur as "one-off" activities. The project manager can use recurrent elements to enhance the accuracy of the forecast due to the reduced uncertainty of the estimates. Asserting the duration of a nonrecurrent task as a single value implies extensive foreknowledge. Describing the task duration as a range of possibilities reflects the uncertainty of project execution. The program evaluation and review technique (PERT) uses a network analysis based on events defined within the project and addresses one-off durations; it allows the project team to express durations as a span of likelihoods. The U.S. DoD classifies estimates as pessimistic, optimistic, and probable. The team weighs its classifications with the heaviest weight going to the most probable scenario. The PERT equation appears as follows:

$$Duration = [(Pessimistic + 4 * Most\ probable + Optimistic)/6]$$

Note that the formula hints at a potentially unjustified normal distribution around the most probable scenario.

The PERT technique provides a framework for simulation. A software tool (@RISK®) exists that provides simulation capability to Microsoft Project.

The PERT estimation technique also provides the project manager with a glimpse of the uncertainty of the estimates. However, the range of values (Pessimistic–Optimistic) provides a strong indicator of the certainty used by the estimator. The project manager will convert this value into the task variance using the equation below. The larger the task variance, the more uncertain the estimate:

$$TaskVariance = [(Pessimistic - Optimistic)/6]2$$

Variations in the three PERT estimates imply uncertainty. However, if the project manager assumes the estimate of time follows a normal distribution, then he can refine or broaden the estimates. Taking the individual estimates to the one, two, three, or six standard deviations (sigma or σ) spreads the available time and improves the probability that the estimate lies within the range of dates. See the table below:

Table 2.3 Sigma and Probability

1-sigma	68.26%
2-sigma	95.46%
3-sigma	99.73%
6-sigma	99.99+ %

Project Time Estimate

Doc Reg Number:		Project Responsible		Prepared by:		Page ___ 1 ___ of ___		Product Name				
Project Responsible:		Key Date		Est Reg Date (Orig) _____ (Rev) _____								

							Calc. Estimate Value (68%)		Calc. Estimate Value (95%)		Calc. Estimate Value (99.73%)	
Core Team												
WBS Designation	WBS Description	Task Estimate Responsible	Optimistic Estimate (hours)	Most Likely Estimate (hours)	Pessimistic Estimate (hours)	Task Variance (+/–)						
1.1.2.	review hw content	JMQ	5	12	15	2.8	8.6	15.2	5.8	16.9	3.0	19.7
						0.0	0.0	0.0	0.0	0.0	0.0	0.0
							0.0	0.0	0.0	0.0	0.0	0.0
						0.0	0.0	0.0	0.0	0.0	0.0	0.0
						0.0	0.0	0.0	0.0	0.0	0.0	0.0
						0.0	0.0	0.0	0.0	0.0	0.0	0.0

Figure 2.4 Duration estimation technique.

The following Figure 2.4 illustrates the effect of variation.

For a confidence interval of 99.73 percent, the range of possibilities varies from 3 hours to 19.7 hours. Estimates with substantial variation should be removed from the critical path or receive risk mitigation. Critical path dates with high variation represent risky goals. PERT models become complicated because the software must iterate through permutations of the three levels—the more tasks/deliverables, the longer it takes for the model to converge.

2.1.5.5   Critical Path Method (CPM)

We define the critical path as the longest duration path in the network diagram—the longest cumulative, connected, slackless lead-time through the project—which means it represents the *shortest* period of time in which the project can be completed. Those tasks on the critical path remain key to delivering the project. The critical path approach calculates the earliest project finish date. The critical path behaves dynamically and may change during the project. The critical path possesses no slack time (the amount of time an activity can be delayed without delaying the early start date of the next task).

The critical path approach suggests that management of slack becomes crucial to the success of a project. The measurement of slack provides us with a risk indicator. As slack dwindles, the project moves toward collapse.

The critical path approach may focus too much on problems as they arise, and less on preventing potential problems. Modern project management software can calculate the critical path quickly and represent it graphically. Software that calculates multiple critical paths treats the project as a metaproject composed of other projects.

2.1.5.6   Network Diagram in General

For planning purposes, the network diagram becomes more important than the more common Gantt chart seen in software programs that support project management. Mathematically, the network diagram derives from the concept of a *directed graph*.

The failure to properly connect the network diagram is probably the single most common scheduling failure by project managers. We started this chapter with some axioms specific to this problem. If the program manager does not connect the tasks based on dependencies, *A* must complete before *B* can start, then the software will inaccurately represent the critical path (see Figure 2.5). Alternatively, an independent task has no dependencies and the team can execute it *immediately*. If such is not the case, the task is not independent.

Figure 2.6 shows the network diagram for the same pseudoproject we used to show the WBS.

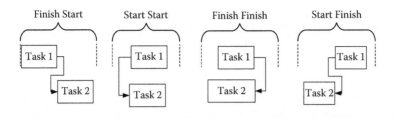

Figure 2.5   Task dependencies.

2.1.5.7   Constructive Cost Model II

Even using the aforementioned techniques, duration estimation is still a *guess* activity. It is possible to develop an association between the activity, the person conducting the activity, and the organization processes. Compiling this information over time allows the project manager or the line organization manager to be able to make some qualifying statements about the validity of the estimates.

Dr. Barry Boehm and a team of others have created mathematical models for just this sort of estimation methodology on a grand scale with a process known as Contructive Cost Model (COCOMO), and COCOMO II.[1] This model is very complex and cannot be adequately handled within a section of a project management book. However, we provide the list below (not exhaustive) to get a perspective on the number of variables that impact the estimation process. Each variable has a number of possibilities or grades. It is no wonder software schedule estimates have accuracy issues.

- Product attributes
  - Required software reliability
  - Size of application code
  - Complexity of the product
- Hardware attributes
  - Performance demands
  - Memory demands
  - Required turnabout time
- Personnel attributes
  - Software team capability
  - Application type experience
  - Programming experience
  - Level of teamwork
- Organization attributes
  - Communications
  - Team distribution (collocated or distributed)
  - Process maturity

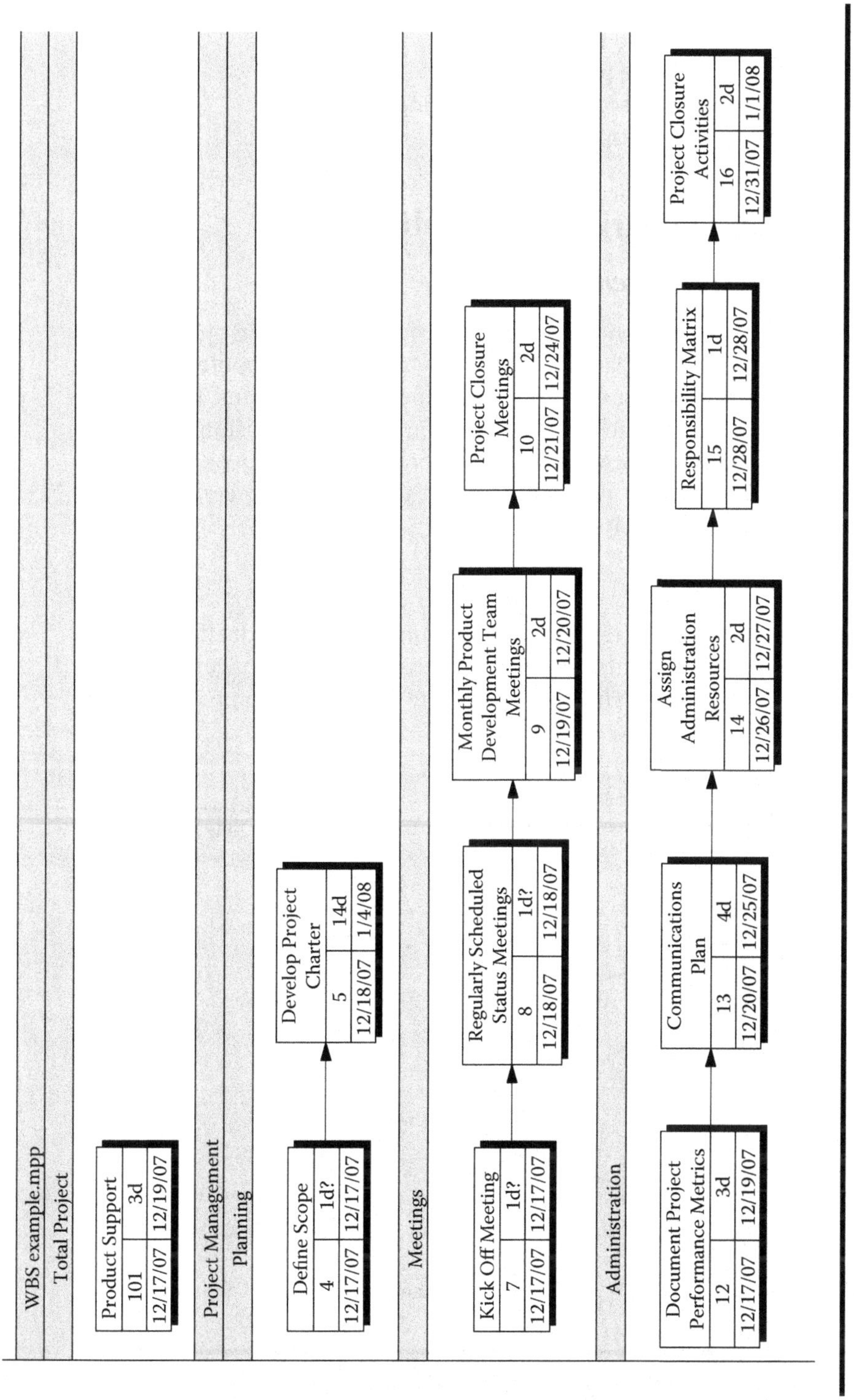

Figure 2.6 Network diagram.

- Project attributes
    - Amount of code reuse
    - Use of software tools
    - Application of software engineering methods
    - Required development schedule

2.2 Product Integrity and Reliability

2.2.1 Risk Management

Figure 2.7 illustrates, in general, the relationship between the project phase risk probability and financial effect. This may seem to run counter to expectations. However, consider the longer the time the project runs, the more is invested in terms of time and money. Further, the more decisions are made and directions taken, the fewer alternatives or solutions are possible. Therefore, while risk may go down as the project progresses, the consequences of those risks have more at stake.

2.2.1.1 Risk Taxonomy

Risk management takes a significant amount of time and effort from a project manager. In fact, from one perspective, project management is the art of risk management. The following brief list shows common internal risk areas:

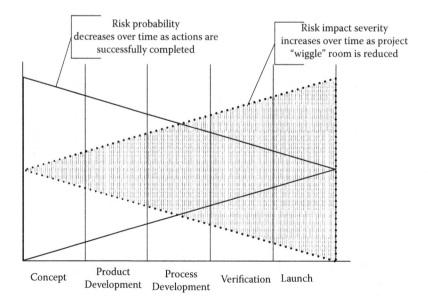

Figure 2.7 Risk probability and effect potential.

1. Engineering
   a. Requirements
      i. Stability
      ii. Completeness
      iii. Clarity
      iv. Validity
      v. Feasibility
   b. Design
      i. Functionality
      ii. Degree of difficulty
      iii. Interfaces to other subsystems
      iv. Performance
      v. Testability
      vi. Hardware constraints
      vii. Software
   c. Coding and testing
      i. Feasibility
      ii. Coding
      iii. Testing efficiency
      iv. Implementation
   d. Integration testing
      i. Test environment (availability)
      ii. Product
      iii. System
   e. Other Disciplines
      i. Maintainability
      ii. Reliability
      iii. Producibility
      iv. Safety
2. Development
   a. Development process
      i. Formality
      ii. Suitability
      iii. Process control
      iv. Familiarity
      v. Product control
   b. Development system
      i. Capacity
      ii. Suitability
      iii. Useability
      iv. Familiarity
      v. Reliability
      vi. System support
      vii. Deliverability

    c. Management process
        i. Planning
       ii. Project organization
      iii. Management experience
      iv. Program interfaces
    d. Management methods
        i. Monitoring
       ii. Personnel management
      iii. Quality assurance
      iv. Configuration management
    e. Work environment
        i. Quality attitude
       ii. Cooperation
      iii. Communication
      iv. Morale
3. Program constraints
    a. Resources
        i. Schedule
       ii. Human resource
      iii. Budget
      iv. Facilities
       v. Equipment
    b. Contract
        i. Type of contract (fixed, etc.)
       ii. Restrictions
      iii. Dependencies
    c. Program interfaces
        i. Customer
       ii. Contractors and subcontractors
      iii. Corporate management
      iv. Vendors
       v. Politics

2.2.1.2 Risk Methodology

We itemize some ways to manage exposure to risk in the list below. The strategy selected depends on the organization and the risk management philosophy.

1. Identify potential risks
2. Analyze risk effect
3. Plan and develop mitigation methods
4. Track or monitor for risk occurrence
5. Control the risk by invoking planned risk response

Risk mitigation is the art of reducing potential effects on the project. Below we show four ways to cope with risk:

1. Risk acceptance – accepting the risk as it matures.
2. Risk transference – assigning the risk to another (the other may be more capable)
3. Risk avoidance – using other strategies to remove the risk
4. Risk mitigation – executing actions to reduce the risk

2.2.1.3 Risk Quantification

A probabilistic concept composed of the following: defines risk

$$Risk = event\ probability \times event\ effect$$

$$Risk = probability \times cost$$

Usually, the estimate of the event occurrence has coarse granularity. However, this kind of preliminary quantification provides managers with enough information to make a decision.

The project manager can estimate multiple risks by multiplying estimates if he assumes independent events. He can look at an example of how this might work. Let's say it becomes necessary to write the specification for the product before a review with key personnel. To achieve the delivery date, he must have the specification written in a specific period *Risk*1 and have the review *Risk*2 within a certain period also.

$$Risktotal = Risk1 \times Risk2$$

$$Risktotal = 0.90 \times 0.90$$

$$Risktotal = 0.81$$

In this example, the probability of achieving the objective of having the specification completed and reviewed amounts to 81 percent.

The project manager can use probabilistic tools such as @RISK and Crystal Ball$^{®}$ to model the project/program using a spreadsheet such as Microsoft Excel$^{®}$ or a project management tool like Microsoft Project. These tools allow the user to run Monte Carlo simulations of the sequences of events and earned value. If the enterprise has a policy of retaining historical data of various projects, the project manager can choose the appropriate distributions to represent various activities in the project (note: not everything follows the so-called "normal distribution"). If he does not know the distributions or knows them poorly, the project manager can estimate some worst-case scenarios and apply a random walk approach to the Monte Carlo simulations by modeling to uniform distributions.

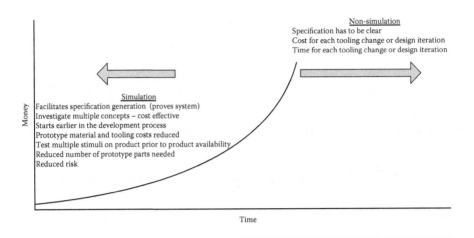

Figure 2.8 Simulation.

2.2.2 Assessment of Product Risk

2.2.2.1 Simulation and Modeling

Simulation makes verification-like activities possible without the material costs. Simulation allows for testing theories and product possibilities without making the actual part. This means it is possible to learn about the product, before much money, time, and opportunity have been sunk into the product (see Figure 2.8). Simulation allows you to adjust the product to better meet customer needs without great tooling costs. However, simulation is only as good as the material of which it is built. The advantages of simulation are great and allow for risk and cost reductions early in the project.

When there are many variations of the system under design, or when the system under design has to interface or is part of a system with many variations, simulation can reduce the logistics around obtaining each of these variations for verification.

There are three types of simulations:[2]

1. **Virtual simulations** represent systems both physically and electronically.
2. **Constructive simulations** represent a system and its employment.
3. **Live simulations** simulated operations with real operators and real equipment.

Virtual simulation Virtual simulation is used to develop requirements by getting feedback on the proposed design solution.

Virtual simulations put the human-in-the-loop. The operator's physical interface with the system is duplicated, and the simulated system is made to perform as if it were the real system. The operator is subjected to an environment that looks, feels, and behaves like the real thing.[2]

Constructive simulation   Constructive simulation is just that, simulating the construction of the proposed solutions. This approach allows quick design changes to be reviewed for impact. Performance information can be distributed to the rest of the team.

Live simulation   Live simulations require the hardware and software to be present. In these simulations, the situations or ambient environment is simulated allowing the system to be checked out under various operational situations. The intent is to put the system, including its operators, through an operational scenario, where some conditions and environments are mimicked to provide a realistic operating situation.[2]

Simulation pitfalls   Simulation and modeling are only as good as the input data. Models must represent the key variables that produce the appropriate systems performance. Additionally, modeling and simulation are specialty knowledge areas. This means the skill set is not often readily available and can be very industry specific. Still, starting earlier, clarifying concepts and requirements means this is a wonderful tool to help produce the product in a timely fashion and at the desired quality.

2.2.2.2   Verification

Any verification of the product, process, or service will provide some data about these products. The project manager must understand that the product, process, or service is a prototype that may not represent the result. However, the purpose of material and process prototypes lies in the reduction of risk to the production of the product or service.

2.2.2.3   Validation

Validation further reduces risk by examining the product or service under more realistic conditions and at a further stage of development. If the embedded team has the software product built, it can model the defect arrival rate with a Rayleigh model and provide the program manager with a statistical basis for final release.

2.2.2.4   Stress Testing

In addition to verification and validation, the team may opt to stress the product or service beyond design limits to characterize performance. Stress testing also yields important data about weak points in the product or service.

2.2.2.5   Reliability Testing

Reliability testing attempts to assess the behavior of the product or service at some specified time. The goal of reliability testing is observable degradation. Under special conditions, the team can model the rate of degradation and predict the life of the product or service.

2.3   Cost

2.3.1   Project Acceptance Criteria

The evaluation of economic gain from a project resembles other corporate investments. The typical criteria are return on investment (ROI), internal rate of return (IRR), or other financial requirements (see Figure 2.9). Some organizations use multiple acceptance criteria, for example IRR and payback period. However, sometimes the enterprise will drive a project for a new strategic relationship with a customer, while not meeting financial expectations. Understanding the rationale for a project allows the project team to comprehend the purpose of the project.

2.3.2   Payback Period

The payback period is the amount of time it takes to return the money spent for the project. If the enterprise spends $100,000 and it makes $20,000 in profit on the product every year, it would take five years to return all of the development monies incurred by the project with the assumption of no effect from inflation. The sooner this payback happens, the more quickly the company makes a profit on the product. Payback period provides a quick means for assessing the cost of a project especially if the payback period calculates to less than one year. Inflation, taxation, and other accounting period-related issues become less significant for short durations.

2.3.3   Return on Investment

Return on investment (ROI) in its simplest form is the ratio of the return or income from the project undertaken to the investment in the project.

$$\frac{Return}{Investment} = \%ROI$$

$$\frac{20,000}{100,000} = 0.2$$

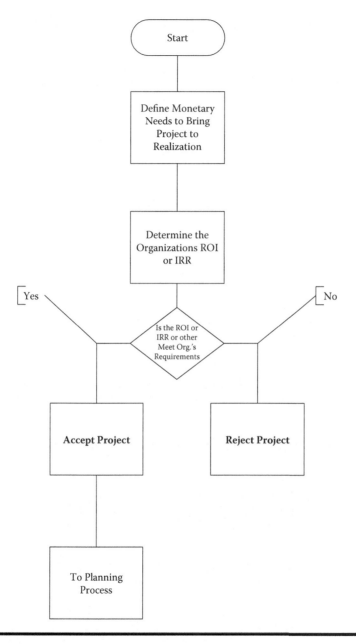

Figure 2.9   Project acceptance.

2.3.4   *Internal Rate of Return (IRR)*

Internal rate of return or IRR, is the annualized compounded return rate for the investment. If a projects rate of return is better then the alternative uses of the funds, the project is deemed to be a good investment and acceptable. Will we make more money investing in this project then another, or even another type of investment (bank or stocks).

$$For \ Internal \ Rate \ of \ Return \ NPV = 0$$

$$-100 + \frac{120}{[(1 + IRR/100)^1]} = 0$$

$$IRR = 20$$

2.3.5   *Market Share*

Sometimes projects are not undertaken for a particular dollar amount, such as a ROI or IRR. Projects can be a useful tactic for grabbing market share from a competitor or for achieving a long-range organizational strategy. The results may be more difficult to quantify than ROI and IRR; however, given the investment, it has tremendous significance for the future of the enterprise. Even when the project evolves into a strategic initiative, the board of directors will normally require a link to the long-term profitability of the enterprise—board-level governance of corporations usually requires a rationale for an initially unprofitable strategy and is frequently a regulatory obligation.

2.4   Project Cost Management

2.4.1   *Earned Value Management*

The cost control procedures define the process interactions and the tasks needed to manage the delivery of the project costs. The team will need to deliver the product and process within the identified cost boundaries. Any change initiatives should also be managed to the same limitations.

To be able to use earned value management (EVM) techniques, the processes and systems in place must at a minimum have the following characteristics:

1. Sufficient breakdown of budget allowing linking of the WBS to the budget
2. Correct billing of hours to tasks
3. Quick response from the hour billing system (latency between when time is put in and when it is visible)

4. Definition of task progress, for example use 0 percent (not started)–50 percent (started)–100 percent (completed) to quantify task disposition

2.4.1.1 Budget Controls

Some organizations do not have tight controls over the hours recorded by employees nor do they have links to the WBS. The actual monetary status of the project may be indeterminate. Only those people working on specific WBS elements should bill hours to those elements. It is just as important that those working on the WBS elements know the accounts to which to bill the effort and do so. Failure to follow these provisos makes it challenging, if not impossible, to use EVM.

EVM arose from U.S. DoD cost accounting and is not unique to automotive development. Project managers use the technique to assess the current cost/schedule status of the project. The tool evaluates the project schedule and cost expenditures against the planned time and cost to determine the status of the project. The system requires detailed preparatory work, most important of which is the WBS.

Let's assume that the project team has identified the scope, tasks, and estimates for the project. The most common name for these variables is planned value since it shows expected expenditures for any given time. Other documents refer to planned value as budgeted cost of work scheduled (BCWS). Once we have the planned value, we can compare it to the actual cost. Other resources may refer to actual cost as actual cost of work performed (ACWP). The time reporting systems have rigid constraints. The project manager must ensure that the people doing the work record their time accurately.

The earned value is the budget at completion (BAC) multiplied by the percentage of completion of the project:

$$EV = BAC * \%Complete$$

Table 2.4 CPI and Interpretations

CPI	Description	Project Status
CPI > 1	The money spent is less than the estimated amount to accomplish	Cost estimates suspect
CPI = 1	The money spent is equal to the estimated amount to accomplish	Project approval
CPI < 1	The money spent is greater than the estimated amount to accomplish	Cost overrun

Table 2.5 SPI and Interpretations

SPI	Description	Project Status
$SPI > 1$	The time to accomplish is less than the estimates	Schedule estimates suspect
$SPI = 1$	The time to accomplish is equal to the estimates	Project approval
$SPI < 1$	The time to accomplish is more than the estimates	Behind schedule

2.4.1.2 Cost Performance Index

The cost performance index (CPI) is the ratio of earned value to the actual cost.

$$CPI = EV/AC$$

2.4.1.3 Schedule Performance Index

The schedule performance index (SPI) is the ratio of the work performed to the value of the work planned. An SPI of 1 means the project executes as planned.

Example: We plan four weeks to execute a given set of tasks and constrain planned cost to $16,000. After two weeks of work, we accomplish 25 percent or $4,000 of the task

$$SPI = EV/PV$$

$$SPI = \$4,000/\$8,000$$

$$SPI = .5$$

2.4.1.4 Cost Variance (CV)

Cost variance (CV) is the dollar amount difference between actual spending and planned spending at specific points in the project. The calculation provides quick feedback on whether the project spending occurs according to plan.

$$CV = EV - AC$$

Example: A certain set of tasks was budgeted to cost $4,000. When the tasks were accomplished, the money spent was $6,000.

$$CV = EV - AC$$

$$CV = \$4,000 - \$6,000$$

$$CV = -\$2,000$$

Table 2.6 CV and Interpretations

CV	Description	Project Status
CV > 1	The amount of money spent is less than the budget	Budget estimates suspect
CV = 1	The amount of money spent is equal to the budget	Budget approval
CV < 1	The amount of money spent is more than the budget	Under-budget

This means that the secured budget for this project is in trouble. There is a shortfall for this set of tasks that may perturb the remainder of the project.

2.4.1.5 Schedule Variance (SV)

Schedule variance is much like cost variance in concept; however, in this case the dollar amount represents the specific amount spent in relation to the project schedule.

$$SV = EV - PVR$$

2.4.1.6 Estimate at Completion

The project manager ensures that the stakeholders understand the project status. This includes informing those stakeholders whether the present budget to complete the project is profitable and elucidating any significant trends.

$$\$EAC = \frac{AC}{\%Completed}$$

Table 2.7 SV and Interpretations

SV	Description	Project Status
SV > 1	The amount of time to accomplish is less than the allotted time	Schedule estimates suspect
SV = 1	The amount of time to accomplish is equal to the allotted time	On schedule
SV < 1	The amount of time to accomplish is more than the allotted time	Behind schedule

Example: A project is budgeted to cost $200,000. It is not at the 20 percent completion mark and has spent $60,000.

$$\$EAC = \frac{\$60,000}{20\%}$$

$$\$EAC = \$300,000$$

This simple equation provides a back-of-the-envelope check to see if the program is on/over/under budget. Clearly, the project is in trouble.

2.4.1.7 Estimate to Complete

The amount of money needed to complete the project from the previous calculated project example is

$$\$ETC = EAC - AC$$

$$\$ETC = \$300,000 - \$60,000$$

$$\$ETC = \$240,000$$

2.4.1.8 Variance at Completion

Variance at completion (VAC) provides the dollar amount of the difference between what was originally planned to accomplish the project to new realities discovered as a result of project execution.

$$\$VAC = BAC - EAC$$

$$VAC = \$200,000 - \$300,000$$

$$VAC = -\$100,000$$

Now, the example project will require an additional $100,000 to complete, if nothing else changes (for example, scope or feature reduction).

2.4.2 Quality, Function, Deployment, and Cost Status

The project team may create custom tools due to pressures from within the project, a simple matter of creativity matching needs. Figure 2.10 illustrates the key tasks by project phase. The figure does not present an exhaustive list but, rather, core tasks identified in the automotive industry action group (AIAG) advanced product quality planning (APQP).

Project Number

Activity	Status				Q	D	C	F	TOTAL
	Yes	No	Late	Inadequate					
	100	0	50	30					
Program Definition									
Quality Function Deployment Activities					75			25	0
Preliminary Engineering Bill of Materials					25			75	0
Product Specifications							25	75	0
Specification Reviews					75			25	0
Establish Quality and Reliability Targets					100				0
Product Assurance Plan					100				0
Preliminary Manufacturing Bill of Materials							50		50	0
Preliminary Process Flow Diagram					25	75			0
Special Process Characteristics					75	25			0
Software Quality Assurance Plan					75			25	0
				QDCF Sum for Next Gate					0
Product Development									
DFMEA					75			25	0
DFMA					50		25	25	0
Key Product Characteristics Identified					50			50	0
Design Verification Testing					75			25	0
New Equipment and Tooling Requirements					50	50			0
Gauges R&R					50				0
Product Test Equipment					75			25	0
Engineering Bill of Materials Released					75			25	0
				QDCF Sum for Next Gate					0
Process Development									
PFMEA					75	25			0
Key Control Characteristics					100				0
Process Control Plan					75			25	0
Special Process Characteristics									0
Process Flow					50	50			0
Process Floor Plan					50		50		0
Pre-launch Control Plan					75	25			0
Process Instructions					100				0
Process Verification					75			25	0
Product and Process Quality System Review					100				0
Measurement Systems Analysis									0
Packaging Specification					50				0
Packaging Specification Review					75	25			0
Process Capability Study					100				0
EDI							25	75	0
				QDCF Sum for Next Gate					0
Validation of Product and Process									
Design Validation Plan and Report (DVP&R)					75			25	0
Preliminary Process Capability								100	0
Bench Testing								100	0
Systems Testing								100	0
Measurement Systems Evaluation					100				0
Production Part Approval								100	0
Packaging Evaluation								100	0
Production Control Plan					50	50			0
				QDCF Sum for Next Gate					0
Release to Production									
Process Sign off					100				0
Trial Production Run							50		50	0
Pilot Runs					100				0
Run at Rate					50	50			0
Production Test Equipment Evaluation					75	25			0
Design Change Notification					50			50	0
				QDCF Sum for Next Gate					0
				Weighted Value					0

Figure 2.10   Project status.

2.5   War Story

2.5.1   Human Resource

At a critical point in the late stage of developing a new product, a key participant can leave the team or the enterprise. This person may have been responsible for the design of the printed circuit board for a high-profile customer. The board had already been laid out and was on its way to the board fabricator. (Note: this situation occurred before the autolayout feature was available to printed circuit board designers.) When the boards came back from the manufacturing facility, the engineers discovered an error. An argument ensued about whose responsibility it was to verify the printed circuit boards. Rather than finger-pointing, it would be more productive to focus on recovery instead of squabbling about responsibility. The project manager should record incidents like this one so that it can be used for instruction:

1. The program manager needed a contingency plan to handle lost member situations.
2. The subsequent counterproductive arguing added no value to the product or project and had a negative impact upon team morale.
3. The launch or change process had no control point in this portion of the process (control points generally involve inspection of work), so the error propagated through the system.

2.5.2 Systems Thinking

A truck company launched a vehicle with the component outlay as shown below. The square component was a device that converted the signals from one type of data communications to another. In addition, the green component worked with a number of data busses and had the hardware on board the micro-controller to handle the bus interface. Software would be required to make these components work together and omit the $30 module from the vehicle build while retaining the functionality (see Figure 2.11). This redundant and costly situation was avoidable by using a design review at the system or metalevel. Metalevel reviews provide an opportunity for developers of embedded software, services, or manufacturables to reassess their work in light of a higher-order systems approach. The review team assesses components for cooperative behavior. This cost reduction should have been available earlier and would have been less than the cost to develop two different components to meet the functional requirements.

2.5.3 Project Manager Words

In the course of a postproduction cost evaluation and improvement exercise, the project manager of the supplier made statements regarding the cost of a particular component (a liquid crystal diode or LCD) within the product. The customer had identified a supplier of the LCD as having a component of similar quality for much less. The prospect of using a much less expensive component while maintaining the same level of performance and design requirements excited the customer. After a quick investigation

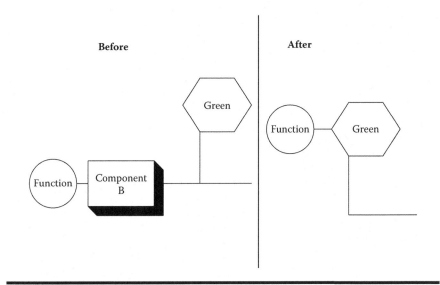

Figure 2.11   Systems view.

of the actual cost by the supplier, the money saved was much less and would not have been cost effective to change within the product considering the cost of product qualification (testing, FMEA review, etc.). In short, any time the project team makes a change, the total cost of change is a significant consideration. Additionally, the team should review the impact of purchased components on the supply chain, given the difficulty of planning for items with weeks- or months-long lead times.

Chapter Notes

[1]Dr. Barry Boehm, Bradford Clark, Ellis Horowitz, Ray Madachy, Richard Shelby, and Chris Westland. April 1995. Software Technology Conference, An Overview of the COCOMO 2.0 Software Cost Model. http://sunset.usc.edu/research/COCOMOII/ (accessed February 16, 2008).

[2]Defense Acquisition University Press, Systems Engineering Fundamentals, (Fort Belvoir, Virginia, Dau 2001) p. 118.

Chapter 3

Concept

The activities where the broadest possible brush paints a picture of the product and processes are the *concept* phase of the program or project (see Figure 3.1). The conceptual approaches identified from the project, process, and product perspective place constraints on future possibilities. Costs and feature limitations grow into powerful constraints on the product and process—a risk that grows during later phases. It is important to understand the nature of these boundary conditions since they confine future movement within the project—motivating prioritization of early project work.

3.1 Concept Overview

The following sections walk through an advanced product quality planning (APQP) project phase for the voice of the customer. Figure 1.2 in Chapter 1 illustrates these phases. In that figure, many of the phases start concurrently (product development and process development). In reality, this presents some risks to the success of the project, since concurrency occurs most commonly with fast-tracked projects. When the team develops the manufacturing process around a design that drifts, the final design result may require risk mitigation that uses sequential phasing of the development of the product and the process.

In this phase the team identifies needs; for example, the launch team researches suppliers and solicits them for concepts using a request for proposals (RFPs). The engineers generate design concepts, possibly using quality function deployments (QFDs) and requirements and constraints techniques. The result becomes a set of ideas that are consistent with the customer's expectations. This situation holds whether the development work comes

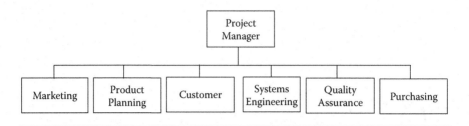

Figure 3.1   Example of voice of customer development team.

from internal or external suppliers. By the end of this phase, the launch team selects one concept and documents it; the launch team also selects suppliers for further work. There may be some early part models in this phase of the project, much like adult show-and-tell. These models usually do not work; however, they provide an illustration of the concept and they resemble the final product. Sometimes these parts receive a designation such as the "A" level in the system of some organizations.

According to the International Council on Systems Engineering (INCOSE), requirements should have the following attributes:

■ Unique identifier
■ Express a need
■ Verifiable
■ Attainable
■ Clear

Suppliers and customers distinguish binding requirements from non-binding requirements by the use of specifically defined words such as "must," "shall," and "will." Specifications can use other words, including "may," "might," and "can" to describe anticipated design characteristics, but they do not impose binding or additional requirements.

The word "shall" states binding requirements of the system/subsystem defined by this document—this usage is common in U.S. government specifications. These requirements will later require verification and validation through testing.

The word "will" states either of the following:

■ Conditions that result from the immutable laws of physics
■ Conditions that result from adherence to other stated binding requirements.

Suppliers can mitigate the risks inherent in the product estimation process by using preexisting hardware and software modules. The performance

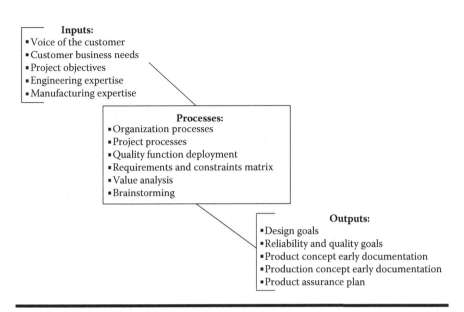

Inputs:
- Voice of the customer
- Customer business needs
- Project objectives
- Engineering expertise
- Manufacturing expertise

Processes:
- Organization processes
- Project processes
- Quality function deployment
- Requirements and constraints matrix
- Value analysis
- Brainstorming

Outputs:
- Design goals
- Reliability and quality goals
- Product concept early documentation
- Production concept early documentation
- Product assurance plan

Figure 3.2   Voice of the customer.

of past projects from the supplier and some technical judgement and experience with similar projects can help mitigate risk.

During this phase, the team clarifies the scope of the project in great detail—an example of the typical team construction appears in Figure 3.1. Dialogue with customers characterizes this phase. Clarification and understanding of the design objectives are the goals of this phase and selection of a concept to achieve the project targets to include preliminary cost, functionality, and delivery schedule. This sets the course for pursuing the objectives and achieving them (see Figure 3.2). The economic merits of the idea make the business case. If the team generates minimal ideas with low probability of target achievement, then the project may terminate or require stronger ideas.

During this phase, the team has the most flexibility in costing. The team gathers, reviews, and issues preliminary plans describing the options for meeting requirements.

The team may consider the program definition phase completed when it selects the suppliers and has satisfied all formal documentation requirements.

- ■ Initial documentation
- ■ Concepts generated
- ■ Supplier detailed documentation and bids (estimates)
- ■ Concept selection
- ■ Supplier award

3.1.1 Inception

The inception phase identifies what constitutes *success* for this project. This phase provides the context for setting the project scope. In this instance, identifying a concept and the production of specifications to support product and manufacturing requirements are typical for this phase. Interacting with the customer—which can include contact with the customer's customer—properly goes into the construction of these product and manufacturing specifications. Further, the team will need manufacturing expertise to produce the relevant documentation for fabricating the product. The team, sometimes in concert with marketing and/or accounting departments, quantifies cost estimates for the project and the product. They endeavor to identify all project deliverables. In addition, they will categorize all quality assurance activities.

The team reports estimates of cost and time to fulfill the requirements of the project. These estimates exist not only for this phase but for the entire project. The estimates grow less valid for the long term than for the near term, which means phase estimation accuracy for the current phase has a higher probability of accuracy than the estimate for the entire project. As the team moves into later phases, project estimation accuracy increases.

Typical scope of the voice of customer phase is:

- Early project scope
  - Cost targets
  - Schedule targets
  - Quality targets
- Business case
  - Estimate of selling price of product
  - Product market volume
  - Estimate of cost of product including organizational cost to deliver to market
- Early product scope
  - Product requirements (functions)
  - Production requirements
  - Reliability requirements

Every process has a beginning, sometimes called "commencement" or "inception." Once started, the process will have inputs processed into outputs; in short, a transfer function drives inputs into outputs.

3.1.2 Planning

During planning, the team identifies all tasks, deliverables, and activities to produce the scope of the project. This includes the hardware deliverables (for example, prototype product) and the quality assurance activities such as reviews. There must be a plan to achieve all contractual obligations to the customer.

The inputs are

- Voice of the customer
- Profitability requirements (customer and supplier)
- Business and market requirements
- Regulatory requirements
- Scope of project (from inception process)
- Delivery desires
- Budget, cost, and timing
- Quality assurance activities

Generally, the project manager prepares project plans and processes and articulates to the entire project team. These plans and processes apply to all phases.

- Document change and revision control
- Risk management plan
- Organizational plan (human resources)
- Communications plan
- Project plan (schedule)
- Cost management plan
- Time management plan
- Quality assurance plan
- Procurement management plan
- Project integration management

During the planning phase, the team and the project manager identify project metrics for use in the regulating operation. Examples of some project metrics for this phase are

- Earned value management
- Number of concepts generated
- Specification status (not started, draft, complete)
- Specification reviews

In the planning process, the project team considers how to meet the objectives defined in the inception phase. The team focuses its planning around the deliverables identified in the inception phase and the acceptance criteria. Additionally, risk management begins with the planning process.

- Identify the risks
- Assess or quantify the risk

- Plan response to risk (plan around the risk)
- Document the risk response
- Monitor the risk
- Lessons learned in the risk's processes

The project manager and the team set about setting and documenting project operating constraints. By this, we mean change management (engineering and project), error, and fault reporting methods. This statement does not imply there are no changes in these processes during project execution—before interaction and execution, all participants must know these ground rules.

3.1.3 Implementation

The implementation process defines the activities performed on the deliverables (outputs) for this phase.

This process produces the deliverables defined in the inception process and planned for in the planning process. In the end, these activities produce deliverables. The automotive industry action group (AIAG) project management guide shows deliverables, provided in the list below, as output from this project phase.[1] The rest of the chapter describes the activities typically undertaken to produce the deliverables. Note that not all deliverable items are hard deliverables such as the product itself. In this case, since the project team listens to the voice of the customer, the deliverables include product documentation and concept selection. Deliverables can also include intermediate items such as market studies or quality assurance.

- Design goals
- Reliability and quality goals
- Preliminary bill of materials
- Preliminary process flowchart
- Preliminary special (key) product and process characteristics
- Product assurance plan
- Management support

3.1.4 Regulation

During the regulation process, the project manager receives project status, performance reports, and quality reports (product status) as an output of the previous phase and determines if the project remains on course. If not, the project manager determines corrective actions and implements them to bring the project back on track and meet the deliverable requirements. These activities include reworking the project plan and placing additional controls on the implementation phase to enable achievement of

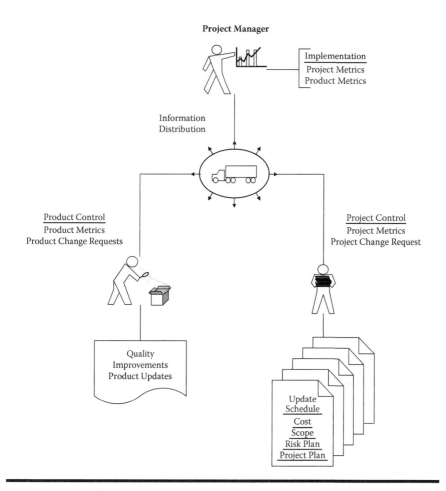

Project Manager

Implementation
Project Metrics
Product Metrics

Information
Distribution

Product Control
Product Metrics
Product Change Requests

Project Control
Project Metrics
Project Change Request

Quality
Improvements
Product Updates

Update
Schedule
Cost
Scope
Risk Plan
Project Plan

Figure 3.3   Regulation process.

the phase targets. During the regulating operation, the project manager receives this information from project and product metrics identified in the
planning operation. See Figure 3.3. In some instances, these metrics come
about as a result of missed project expectations or discovery of probable
risks. However, the team has identified some metrics from previous similar
projects and some remain a part of the project from the start. An example of
applicable metrics for this phase would be the identification of the number
of specifications needed and a means for tracking the quantity and reviewing the amount of time consumed. This monitoring would allow projection
of the completion date and determine if the project achieves the targets
identified in the inception phase. When the delivery dates and expectations remain unmet, the project stakeholder needs to know immediately.

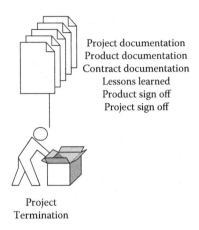

Project documentation
Product documentation
Contract documentation
Lessons learned
Product sign off
Project sign off

Project
Termination

Figure 3.4 Termination.

Sooner is much better than later in order to avoid the appearance of hiding information from customers—internal or external.

3.1.5 Termination

In general, during the termination phase (see Figure 3.4), all documents required by the project are current and, if the project or launch team is dealing with an automotive customer, the appropriate documents readily ship to the customer as a production part approval process (PPAP) submission. All other deliverables (especially the product or service itself) are also current. If the deliverables are not at the required level of quality, then the project is at risk and management may reject closure of the project. The team should assess risk levels to the project and what, if any, actions need to occur in order to mitigate these late-in-this-part-of-the-process risks.

The termination process compares the deliverables targeted or specified in the beginning of the project, including the quality requirements, to the final deliverables. In this particular phase, the end product should be a concept that meets the customer requirements with appropriate evidence (for example, a preliminary design).

The team and management set up the termination process acceptance criteria during the inception process of the project. They use the planning phase to identify methods of achieving goals. If they omit a requirement, then this phase will not close or exit without stakeholder buy-in and any subsequent phase would be at risk.

The termination process completes the vendor activity and can also serve as a time for review.

According to the *AIAG Advanced Product Quality Planning and Control Plan*[1] the outputs from the voice at the end of the customer phase are:

- Design goals
- Reliability and quality goals
- Preliminary bill of materials
- Preliminary process flowchart
- Preliminary special (key) product and process characteristics
- Product assurance plan
- Management support

3.1.5.1   Gate Targets—Past View

- Design documentation quality secured
- Project, quality, cost, and delivery objectives met
- Development cost to date compared to projections
- All contractual obligations met

3.1.5.2   Gate Targets—Future View

- Need for developed solution still exists
- Margins on design still fit organization's needs
- Future risks not increased since start
- Estimates to deliver the remaining portion of the project still meet business objectives

3.2   Delivery

3.2.1   Work Descriptions

3.2.1.1   Project Charter

The project charter facilitates project overview for participants and stakeholders. This device communicates key project aspects. The charter does not replace other formal documentation, but distills key project information from many formal documents (for example, the project schedule) into one document. It establishes the high-level goals for the project, not just in terms of monetary success, but as a description of other less tangible goals of the project such as customer satisfaction.

This project quick reference helps focus the project team when it selects the appropriate elements. An effective project charter communicates to the customer and to the team members the scope of the project.

Charter elements   The elements of the project charter vary depending on the needs of the particular project, project manager, or to meet specific organizational or customer requirements. The primary content should derive from the specifics of the project.

The template content should reflect the priorities of the organization. Generally, after working to establish the organizational requirements for the charter, the requirements themselves and the associated documentation become a template for other projects. The identified elements should support the processes and the philosophy of the organization.

In the end, this document becomes a quick reference for key issues within the project and a way to keep the participants focused on requirements. Below is an example of possible elements for a project charter.

Project mission   In the places we have worked, mission statements have a bad reputation. They become fodder for many *Dilbert*® cartoons and other management spoof books. However, when the team and project manager craft a mission statement by sharpening and defining the project team focus, it becomes a valuable tool.

Focusing on the stakeholder and the desired outcomes of the project requires a great deal of effort. All of this unnecessary or redundant reframing of the mission emerges as a collection of nonvalue-added activities. That is not to say the project manager and team have wasted time spent clarifying the expectations and the mission. Even when the project manager and team create a project charter and a mission statement, project success is not assured. Repeated reviews of the mission by the team instills the mission into the team.

A good project mission statement facilitates achievement of objectives. The important aspect of the mission statement lies in the engagement of the team during the delineation of the mission.

Scope   The importance of scope cannot be overstated. Without a clear and precise comprehension of the definition of the project target, achieving that target is largely luck with a small measure of skill. If the team cannot articulate the target, then it will not get the project participants to unite to meet the target. We call this management by hope, with little or no chance of being successful.

Generally, we define scope to consist of timing, quality, cost, and feature content. There are many illustrations of scope as a three-dimensional triangle, with the sides composed of timing, cost, and feature content. The scope includes specific project deliverables.

Scope creep   This happens to many projects. This situation is not just an automotive notion, but often a major cause for a project to not deliver on time or at the original cost estimates. If the project manager does not ensure adequate time spent on determining the scope of the project, he may feel the affliction of scope creep. If the project deliverables are not well known, the project has opportunity for scope creep. With scope creep come budget overruns and time slippage.

Does our critique mean scope creep is always a negative thing? No, there are times when a last minute requirement makes it to the development team because it is necessary or a response to a change in the reason that precipitated the project. It is important that this change to the scope be considered vis-à-vis the effect on schedule, delivery, or quality. The effect on the budget and the variables being measured—cost, performance index, schedule performance index, and the other earned value calculations—may not reflect the actual performance of the project if the project manager does not control the expansion of scope. Occasionally, this additional scope can be absorbed by the project with additional resources. No matter how small the change, a thorough planning of how the team will execute the additional requirements and the effect on customer expectations must be understood and communicated back to the customer. There are times when scope creep is a response to business opportunity and a quick turnaround of the feature ensures project success or improved market penetration. There are certain characteristics of scope creep that are important. Scope creep is

■ Often initiated by the customer,
■ Difficult to detect until an adverse or catastrophic effect has been encountered by the project,
■ In the customer's best interests.

The project scope must be managed to minimize and eliminate the negative effects of scope creep. That is done by

■ A well-defined contract that is supported by a detailed work breakdown structure,
■ A change management plan for the project,
■ An endorsement obtained from management,
■ Validation for the change.

Scope creep can become a source of ready revenue for the supplier (services, embedded software, or product). The key takeaway of this section lies in maintaining control and estimating impact (consequences) when the customer desires a scope change.

During project execution, deviations from the scope and contract *will* occur. A key to customer satisfaction is that the customer does not suffer from surprises. Quality creep can be considered a subset of scope creep. To avoid quality creep, use metrics and remember the difference between expectations and needs. It is necessary to acknowledge the risk to the project of any change to the scope. Some changes create greater risks than others. It is important to assess the risk implications to any change; for example,

- Expensive decisions
- Limited budgets
- Costly external forces
- Immediate cost vs. life cycle.

It is important that a system for scope containment and management is identified early in the project. It should be considered during the contract creation phase of the project. The contract provides the baseline for the project scope. The marketing/legal staff should detail the contract enough to provide quantifiable criteria for which the project team is accountable. Also needed is an acceptable change management process to ensure that the changes encountered during the execution of the project receive expedient handling with control over the potential effect on the project. At a minimum, the change management plan should

- Identify the change type and source;
- Analyze the effects of the change cost, quality, and schedule;
- Develop a response strategy and assign responsibility;
- Communicate the strategy and gain endorsement for the change;
- Revise the work plan and monitor the effects of the change.

Team identification, responsibility, and availability This section identifies the team members and their roles and responsibilities within the project. This section is not meant to provide a detailed breakdown of member responsibilities, but to provide an understanding of the project structure, overview of areas of influence and percent of time spent allocated to the project.

Identify customer and stakeholders These people have an interest in the outcome of the project. From an internal perspective, this would be the project sponsor and other parts of the organization such as production for the resultant development project.

The customer is the most obvious and still often overlooked stakeholder. Identifying the specific customer representatives by name improves the likelihood of managing and meeting their expectations. While identifying stakeholders does not guarantee success, it is a sure recipe for failure to not identify them. With stakeholder identification, the team then knows who expects results from the project—it puts a face on the customer. Distributing this information to the project team presents the opportunity to better understand stakeholder expectations. If no one takes time to identify their concerns, the success of the project diminishes.

Stakeholders and expectations Stakeholder expectations link to the project scope. Seldom does the project team have stakeholder consensus about project success at project kickoff. The project team (particularly the project manager) must understand these expectations and clearly convey those expectations to the rest of the project team, including the launch team, the embedded development team, and whoever else participates in the project. The project charter is one excellent tool to enable focus by the project team.

There are two types of expected project outcomes: measurable and immeasurable outcomes. While both are important, the most difficult for engineers to understand are unqualifiable outcomes, those that are immeasurable per se.

Anticipated project outcomes—measurable At this point in the process, the team has identified stakeholder expectations. Use of stakeholder expectations allows the team to devise a plan of how the project is going to match stakeholder expectations. The team can also use stakeholder expectations to drive the measurement plan. The selected measurements and related thresholds should support and verify the degree that expectations meet some boundary value. A plan or strategy that does not support stakeholder expectations means the project success rests on management of the stakeholder expectations (perception) instead of guiding project progress toward the expectations *and* managing those expectations. To achieve this level of guidance, key questions must be asked:

- What are stakeholder expectations?
- How can those expectations be quantified?
- How can those expectations be qualified?
- How do we ensure we meet those expectations?
- When will we know that we have drifted away from stakeholder objectives?

These questions need to be answered to define what constitutes project success. It involves understanding what is expected and a way of measuring if the objective or outcome was met. Examples of measurable project outcomes are

- Project costs
- Designed component piece cost
- Schedule

Anticipated project outcomes-immeasurable The immeasurable outcomes are more difficult to quantify but are often of equal or greater importance. Occasionally, the project team provides the appropriate deliverable products, but alienates the customer with costing pettifoggery, variable quality, and unpredictable delivery times in the course of exercising the project. The customer or stakeholder may decide the product is not worth the effort put into the process. While we have not found an objective measure for customer irritation, we may be able to infer their perception by observing responses to communications (especially phone calls), delays in payment, and other petty behaviors that suggest they are suffering displeasure with the project.

Major milestones dates Shrewdness of project participants regarding key project dates is vital to project execution. The more the members of the team know of key project milestones, the better likelihood that the risk of meeting those objectives will be presented to the team and its leadership. Such key milestones may include, but are not limited to

- Start of project
- End of design phase
- Design validation testing
- Start of production phase
- Process validation testing
- PPAP
- Production start
- End of project

In the automotive world, the PPAP functions as a key documentary milestone. Some enterprises treat it as a necessary evil; however, we suggest it is frequently the last opportunity a project/launch team has to check up on its own performance before delivering the final product or service.

Project team scope of decisions During many projects, situations arise when the project participants have questions regarding decision boundaries. This circumstance is especially true in an environment trending toward empowerment. Empowerment is not a credit card for making decisions, but moves the decision-making ownership closest to the actual issue.

The project participants must be clear about the areas on which they make decisions. Matching available skill sets to the required tasks generates a responsibility list. This area of responsibility for each of the participants defines the resource scope for each individual.

Some other possible content Charter contents vary from company to company and project to project. The key areas defined by discussions and negotiations with the customer and project planning result reflect the content, some of which might be

- Project phase dates
- Project phase budget overview

- Tooling budgets
- Other project constraints
- Prototype expectations
- Prototype parts and vehicle support
- Initial quality targets
- Penalties for late deliveries

Sign-off   This is the metaphorical dotted line. This part of the charter process captures the commitment of the key participants. The yield of planning and negotiations is a document that stakeholders can endorse. Any changes in the project constraints should receive approval from these individuals. This scenario works to the favor of the supplier also, since changes to the scope or delivery dates are captured in this overview with any proposed changes to the project requiring replanning.

3.2.1.2 *Statement of Work and Statement of Objectives*

Statements of work (SOW) specify the activities required during the course of a project or program. Most often, the customer—internal or external—will prepare the SOW. It is a tool for defining what is in scope and clarifying what is out of scope. A typical SOW might have the following structure:

1. Scope
2. Applicable documents
   a. Department of Defense specifications
   b. Department of Defense standards
   c. Other publications
3. Requirements
   a. General requirements
   b. Technical objectives and goals
   c. Specific requirements
      i. Contractor services
      ii. Integrated logistics support
      iii. Management systems requirements
      iv. Production planning for phase II
      v. Reliability program
      vi. Maintainability program

In essence, the SOW is what the name implies—a clear, intelligible specification of the work it will take to develop and produce some goods or services. The SOW defines the work but it does *not* define the product or service. The SOW is not expected to describe how, but, rather, to specify the results needed to complete the mission.

3.2.1.3 SOW Detail

Scope   As with the project charter, the SOW also describes what is expected and, in some cases, what should *not* be done. Occasionally, introductory and descriptive material may be appropriate as long as the players understand the constraints.

Applicable documents   The customer spells out the appropriate and guiding documents, particularly revisions. If a large number of related documents exist, management of the document tree can become a burden. If some documents function as handbooks or support material, but not as requirements documents, the supplier should request clarification. Other trouble spots can occur when the downstream customer specifies standards that are no longer current. The wise supplier will verify that the obsolete standards are what the customer desires.

Requirements   Requirements are the meat of the SOW and generally have these qualities:

- Clarity
- Specific duties
- Sufficient detail that duties are unambiguous
- Refer to a minimum of related specifications and handbooks
- General information is differentiated from specific requirements
- How-tos are avoided

Work breakdown structure (WBS)   The WBS is a tool to help guide the construction of the SOW. However, the SOW need not go into the level of detail that normally is obtained with a WBS.

If the customer has his or her own staged development process, it may be appropriate to define the stages for the suppliers so that the suppliers may align their own launch process with that of their customer.

SOO detail   The statement of objectives (SOO) is a short document that can provide further clarification of a project by explaining the purpose of the product or service.

Program objectives   This section describes the phasing of the program with an explanation of the phasing system when necessary. It may also constrain the number of tiers of suppliers involved in the development and final delivery of the product.

Contract objectives   This section elaborates on the previous section and explains the purpose of the various phases and subphases. An understanding of the customer system should lead to a more credible alignment of the supplier's development/launch system. If the contract is large and a specific supplier works on a subcontract, knowledge of the overall contract can clarify the *goals* of the project.

3.2.1.4 Charters versus SOWs

A program charter is typically a short document that lists the main highlights of a program. Most commonly, the program manager will author the document and submit it for management oversight. A good charter should provide information about Kipling's six friends: who does the work, what the work is, when the work should be done, where the work will be done and tested, how the work is done (resources not instructions), and some basis for doing the work (the "why").

A good SOW is much more detailed than a typical program charter. A well-written document, authored by the project manager, the customer, or both, can reduce customer and project risk. The downside lies in the potential for micromanagement by the customer, which can increase cost and cause delays. Note also that the SOW will deal with issues like quality, reliability, testing, and other activities that most program charters do not address.

The good SOO probably lies somewhere between the SOW and the program charter in complexity. It is always authored by the customer. An SOO attempts to explain the goals and objectives of the customer in a more general sense without becoming so nebulous as to have no value. For example, an overarching goal might be something like "design and develop a drive train for a medium-duty truck that will increase safety while decreasing weight and cost." Other objectives would then be derived from this statement while avoiding the "how" part of the deal.

3.3 Product Integrity and Reliability

3.3.1 Voice of the Customer

The voice of the customer has a higher priority than the voice of the engineer (or service designer) when it comes to customer satisfaction. Additionally, paying attention to customer desires and needs can lead to profitable long-term commercial relationships. This phase takes input from the customer, regulatory, and business requirements and outputs specifications and project documentation.

The activities undertaken in this phase are the same whether developing for internal enterprise use or for a customer. Ultimately, understanding fulfillment needs and profit level counts a lot—it is true whether with customer or supplier.

3.3.1.1 Quality Function Deployment

Quality function deployment (QFD) facilitates translation of customer desires into the design characteristics: hardware, software, or for a service. As the name suggests, QFD supports two aspects:

1. Quality deployment, or the translation of the customer requirements into product design requirements.
2. Functional requirements, or the translation of the design requirements into the required part, process, and production requirements.

An integrated QFD (see Figure 3.5) process provides numerous benefits to an organization, the foremost of which is increasing the probability of meeting the customer's requirements. Further, when executed properly, it can reduce the number of engineering change requests (ECR) due to increased engineering knowledge of specific customer requirements before

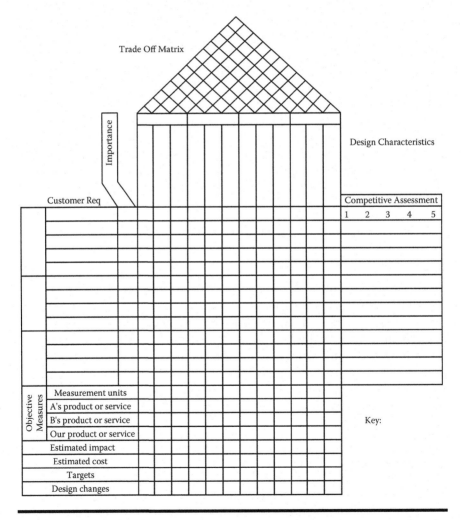

Figure 3.5   QFD example.

the actual design work starts. The presence of numerous ECRs means a lower likelihood of meeting the project budgets and delivery time. The QFD format shown is easy to replicate in a spreadsheet.

A thorough QFD identifies conflicting shareholder or customer requirements. These conflicting requirements can be difficult to resolve once the design work progresses. The situation only gets more difficult as the design moves closer to production.

Ensuring a common understanding of the objectives and functional requirements for a product means less time expressing and negotiating during the delivery phase of the project when the focus must be on achieving design requirements, not understanding and negotiating those requirements.

A QFD is a matrix-oriented tool for defining the *voice of the customer* during the development process. In each case, an important category of information is related to another category of information. This model follows a pattern:

1. Customer expectations answered by design requirements,
2. Design requirements met by part characteristics,
3. Part characteristics handled by key process operations,
4. Key process operations implemented by production requirements.

3.3.1.2   Requirements and Constraints Matrix

Customer specifications may contain constraints on the task (for example, customer-specified supplier). Constraints provide boundaries for the design, such as current consumption, environmental survivability (mounted on frame rail), or regulatory requirements (vehicle example).

One technique for understanding the design requirements is through the use of a requirements and constraints matrix. This technique can be used to illustrate the boundaries for the product.

The juxtaposition of the requirements with the constraints clarifies potential conflicts among and between them. Once the team uncovers these conflicts, the design team can make an assessment about the effect on the requirements and the constraints. If the design will not occur because of these conflicts, the design team can use the analysis to make the case for an alteration to the requirements or to the constraints.

3.3.2   Engineering Issues

3.3.2.1   Bills of Material (Engineering)

Engineering bills of material (BOM) are typically structured by function and are generally the first collection of data that the corporate purchasing department uses to assess the probable cost of the product. Most commonly,

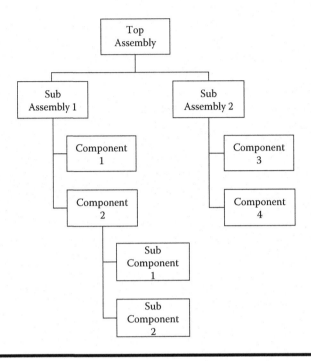

Figure 3.6   Bill of material example.

the engineering BOM is constructed using a computer spreadsheet and represented in a tree format (Figure 3.6).

If the product is embedded, the BOM may consist of a microcontroller/microprocessor and associated code, making a subassembly. If the product is a service, a bill of resources may be more appropriate.

Service designers are more likely to concern themselves with bills of resources than with bills of material. A bill of resources is a document that specifies the resources, often people, needed to achieve the consummation of the service.

3.3.2.2   Change Management (Engineering)

All designs are at risk of requiring changes or engineering change requests (ECRs). Some possibilities are legal changes, late customer requirements, and survivability concerns that arise during the end game. The engineering change processes (see Figure 3.7) must have a way to track these changes, assess the effects of the change on the product and the project, and communicate those results throughout the acquisition and customer organizations. The system for initiating change must accommodate the ability to change

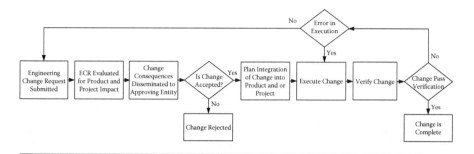

Figure 3.7   Change management process.

the design from either the supplier or the customer, and not allow the change to happen unless agreed on by all required parties.

Most often, engineering changes originate from the customer. However a reciprocating change management system makes it possible for the supplier to propose cost improvements to the design. Whether the change is initiated from the supplier or from the customer, a detailed and common understanding of the requirements of the change and subsequent effect must be known and agreed to by both parties. There should be a formal sign-off and a mechanism for updating the existing product documentation or specifications. Once the required technical change is understood, the effect is evaluated for affect, and the team either accepts or rejects the change. While an enterprise can find a number of tools and specialized systems for tracking these changes, a simple and time honored method is illustrated in Figure 3.8.

Uncontrolled product and project changes have a serious negative effect on project completion. In our experience, these situations happen often and the consequences of *absorbing* the additional scope often have a catastrophic consequence on the project. There must be a formal change management system in place and it must be followed without exception—otherwise placing the product and the project in jeopardy. The costs of making any change increase the risk as the project progresses (see Figure 3.9).

Excel is one tool for tracking the changes; however, a more sophisticated system might be Rational's Clear Quest®, which performs this function by design.

Below is an example of information contained within an ECR.

- Project name/number
- Name of issuer
- Unique change request identifier (per change)
- Date
  - Change issue date
  - Change desired availability date

Project name	**ENGINEERING CHANGE REQUEST**	Order nr:
Customer:		Request nr:

Description and reason for modification request:

Issued by:		Signature	Date:
X Customer employee name			

Effected parts:		
Part:	Part description:	Description of modification:

Consequences to		Description/Calculation	Amount
Product cost	Yes / No	(*Add quotation*)	
Investments	Yes / No	(*Add quotation*)	
Development cost	Yes / No		
Timing project deliverables	Yes / No		
Other parties	Yes /No		

☐ **Approved**   ☐ **Rejected**

Authorization	TechHawk Corp
Engineering Name Signature　　　　　　Date	Engineering Name Signature　　　　　　Date
Purchasing Name Signature　　　　　　Date	Sales Name Signature　　　　　　Date
Project Manager (engineering order) Name Signature　　　　　　Date	Project Manager Name Signature　　　　　　Date
Authorization Name Signature　　　　　　Date	Authorization Name Signature　　　　　　Date

Figure 3.8   Example of engineering change request.

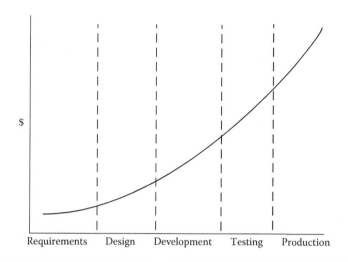

Figure 3.9   Cost of changes to product over time.

- Nature of change
  - Hardware
  - Software
  - Mechanical
- Description of change
  - Reason for change
  - Details of change (specification)
  - Documentation affected (software requirements specification or SRS, hardware specification)
- Change category
  - Enhancement
  - Cost rationalization
  - New feature
  - Supplier initiated
- Priority
  - Low
  - Medium
  - High
- Account number
- Consequences of change
  - Time
  - Cost
  - Quality

3.3.2.3 Change Management (Project Management)

In addition to change management for the physical design or specific engineering aspects, there is a need for a change management system for the project. Project changes include changing the delivery date, the cost, or otherwise altering the scope or imposing other quality constraints different from the original agreements. These types of changes necessitate replanning or recalculation of the project to meet these new delivery requirements. Sometimes, engineering changes create a need for the project to change; for example, a new technology may not work. It is more probable that engineering changes will have an effect on the project than not. For small engineering changes, there may be no effect. However, for any engineering change, there must be an assessment of effect: *every change risks the project cost, quality, and delivery.*

3.3.2.4 Material Specification

If specific material is required to fulfill a design aspect, the team should identify it clearly—by that we mean chemical composition and other characteristics of interest. Specification at this level, however, means a predetermined cost and allows little or no room to achieve a given design requirement cost effectively. This situation is satisfactory if the material specified is the only way to achieve the design objective. Further, due consideration of the manner of handling the material through production is necessary. There may be physical properties that need special handling during manufacture.

The analogous specification for a service might be a subservice. The service designers would specify the required features of the subservice and generate a statement of work for the subcontractor.

3.3.2.5 Engineering Specification

The engineering team derives engineering specifications from the customer functional and performance specifications. Added to these inputs are things the supplier either knows or infers from the customer requirements. For example, the customer may say the new product must live on the *frame rail* of a heavy vehicle. The supplier can infer that environmental test requirements, if not provided by the customer, can be found via SAE J1455 or other standard, particularly if in a non-automotive industry. These requirements are folded into the response to the customer's functional specifications.

The customer specifications can be redolent with relevant details or be based on the goals and performance objectives of the product and not be sufficient to produce the product. When the customer specification contains all the details, we can refer to it as an engineering specification authored by the customer. Many organizations provide such specifications; in fact, it is typically a collaborative result between supplier and customer.

Life testing Being able to respond to the customer's needs means understanding the context of the various requests made to the supplying organization. Understanding the numerous stimuli to the final product—over what can be referred to as the life of the component in the environment for which it was designed—is essential to the quotation and estimation process.

Life testing requires a grasp of the types of environments and exposures a product or component will receive during its life cycle. This information requires time to collect with a sample. This data collection provides information on the exposure of the product to some external stimulus. It is possible to pick the represented environment that would move this measurement from the design limit to the fail or destruct limits. For example, the test team could attack the product with a multitude of overstresses to provide increased verisimilitude and more prompt results.

The service and embedded alternatives to life testing include automated high-speed testing, both combinatorial and stochastic. The goal is to exercise every condition and to provide random stimuli to enhance realism.

3.3.2.6 Customer Specification

Product specification Customer specification exists whether the customer or the supplier writes it. It is essential that both the customer and the various tiers of the supply chain agree on the description and performance requirements of the desired product, be it service or object (see Figure 3.10).

Both the Institute of Electrical and Electronic Engineers (IEEE) and the U.S. Department of Defense (DoD) provide documents that describe formats that can be used to build systematic specifications. Most specifications we have worked with have been written as narrative text with admixtures of tables and figures. In some cases, the narrative text causes more obfuscation than clarification. Since one of the first acts of a responsible engineering organization on the customer side is to break these specifications down into quantitative (or quasi-measurable qualitative) performance requirements, suppliers and customers can consider using a spreadsheet or database format from the start—in a sense, cutting to the chase from inception. Providing a unique identifier for each requirement (alphanumeric designation) identifies the requirements and is a good start to creation of the test plan for verifying that the design meets the requirements during the verification phase. An example is shown in Figure 3.11.

Product requirements are the technical needs and are the basis for achieving the design. The requirements provide insight for the design staff

Figure 3.10   Specification inputs.

to understand the customer. The design can be only as good as the information contained within this documentation.

Detailed requirement documents are necessary to:

- Facilitate project efficiency and delivery (minimizing rework and redesign)
- Ensure only those features needed are created by the design staff
- Allow comparison of the actual design output to the requirements documentation as a way to quantify and verify the delivery of the project.

Note that everything we have covered in this section can apply just as well to any industry including the service industries. The more upfront

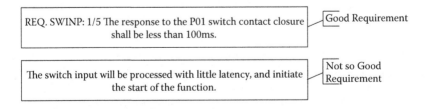

Figure 3.11   Examples of requirements.

work teams can do, the more probable it is that they have stimulated a realistic fault condition, giving themselves the opportunity to eliminate the condition before the customer ever sees it or exercises it.

Service part specification   In almost every project/program we have seen, neither the customer nor the supplier creates a descriptive service part specification. The closest we typically see on a project is a last-minute attempt to create service parts by modifying the drawings for the regular-issue part. While a service part is frequently the same as the regular-issue part—differing mainly in handling and designation—in some cases the part may have distinct differences from the regular-issue part, not the least of which is the packaging. Service parts are more commonly shipped as *onesies* than many parts to a container.

The service part concept is irrelevant to embedded project management unless the customer sees a field condition requiring part replacement (also known as a "recall"). Since the standard service part contains the standard software, the team is unlikely to have to pursue special handling of the embedded software.

For services, one might consider contingency plans for dealing with service failures. The team could consider a higher-level tiger team or some other elite group whose sole purpose is to handle the occasional service failure.

Shipping specification   Shipping specifications are important because they provide for the protection and expeditious movement of the part. Poor packaging design can lead to return merchandise authorization, unnecessary cost, and unhappy customers. Some automotive products (anything with glass, for example) may require special handling. If the lenses of an instrument cluster are polycarbonate, they may require more protection in order to avoid scratches. Wire harnesses required protection for the connectors to avoid damaging the pins.

Shipping specifications are not unique to the automotive industry. Any industry moving hardware will often specify some type of packaging and handling to eliminate damage to the product.

Shipping specifications are irrelevant to services per se with the possible exception of, for example, online services that provide reports.

Customer drawings   Customers do not always supply their drawings to their suppliers, particularly in the automotive industry. It is more common that the supplier provides drawings to the customer for integration into the customer's product management system. The most common interchange format is Initial Graphics Exchange Specification (IGES), which conveys most of the information in the original drawing to a standard translation format. Obviously, it is much more convenient if the supplier and the customer use the same software and version.

Drawings must be under configuration control. In many companies, the drawings are the final word on the product, so they occupy an important place in the product management system. Do not attempt to manage part numbers using the drawing system; this ends up becoming two databases with the corresponding difficulties in maintaining referential integrity.

Changes to drawing part numbers are necessary whenever form, fit, or functional change occurs. It is also common to have part numbers change when there exists the possibility of a quality concern by the change or if the change corrects a possible quality issue. This allows the customer to determine the cutoff date for the affected product, should the possible quality concern come to fruition. In many cases, the cutoff date becomes a so-called "clean date" after which the quality problem should cease to appear.

Product marking   Often, a customer will require that a component or product have special markings. There is a set of frequent requirements:

- Date (month, day, year, or a week–year concatenation) of the manufacture of the component or product
- Recycling information (material composition)
- Serial number for the product
- Company logo (the customer's)
- Revision level of the component
- Manufacturer ID number

If a service involves deliverable items, these can receive serialization also; for example, the required notations on a report.

3.3.2.7   Specification Reviews

Before work is started on the product, the team must review the specification for the product. Even if the specification is the result of a collaborative effort, the team must review it to support common understanding among key participants of the project. The goal is to have the engineers/service providers internalize the customer needs, requirements, language, and terminology.

3.3.2.8   Special Product Characteristics

These are characteristics of the product that are important to the customer and that must appear on the control plan. These product characteristics are those for which anticipated variation will probably affect customer satisfaction with a product; for example, fit, function, mounting or appearance, or the ability to process or build the product.

Embedded code and services have special characteristics also. The concept generalizes well and allows embedded and service teams to account for special customer needs rigorously.

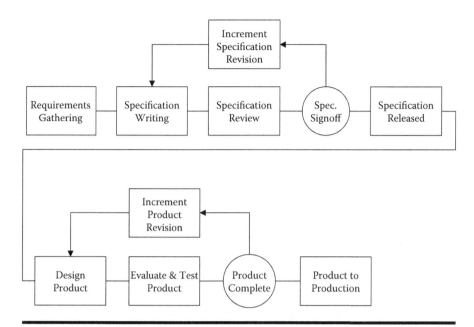

Figure 3.12 Incremental development model.

3.3.2.9 *Hardware Revision Levels*

It is wise to note changes that affect the customer with revision levels.

- Creation: Stage A
- Verification: Stage B
- Tooling: Stage C
- First production: Stage P

In general, we usually see four phases with material being shipped to an engineering facility from the supplier (see Figure 3.12). Often, this material must ship with documentation stating the level of the material (for example, *prototype*) and other relevant information (see Figure 3.13). This documentation is sometimes called a "warrant."

Revision Level A parts   This hardware level is produced early in the creation phase and it may have some other designation than "Level A." This segment of the project includes the concept study and early detailed development phases. The purpose of this stage is to provide early-part information to the customer to get feedback on the design concepts and early-design definition. This preparation allows for early-design improvement comments to translate into productive product changes or modifications before heavy tooling costs happen. During the concept development phase, this feedback allows for comments to be included in the detailed development phase.

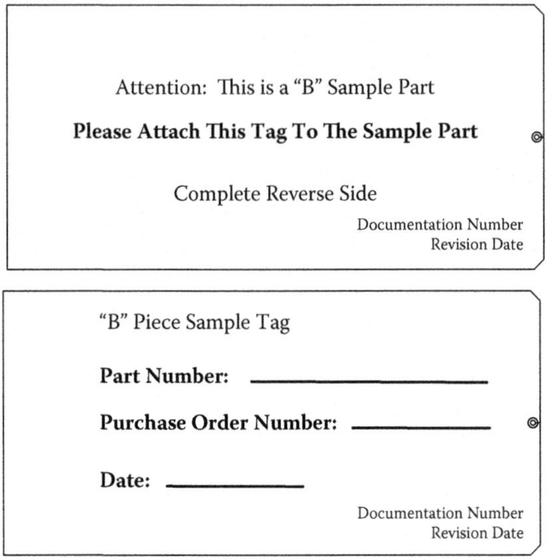

Attention: This is a "B" Sample Part

Please Attach This Tag To The Sample Part

Complete Reverse Side

Documentation Number
Revision Date

"B" Piece Sample Tag

Part Number: ————————

Purchase Order Number: ————

Date: ————

Documentation Number
Revision Date

Figure 3.13   Example of a B part warrant.

Note that embedded software development and services should perform an analog to these parts. With software, the team might see a prototype version that exhibits functionality as proof of concept and as a target for comment from the team and the customer. If they have a *revision level A service*, they can do the same level of feedback and subsequent modification of the service.

Before the first release of the A level of hardware, the following actions should take place:

■ Perform make or buy evaluation
  ■ The enterprise can fabricate the part
  ■ The enterprise can purchase the part
■ Secure part numbers for the proposed parts
■ Complete digital mock-ups
■ Prepare draft documentation of the requirements for auditing and assessment

After the first release of the A level of hardware, the following actions should take place:

■ Identify possible suppliers (if multiple suppliers of the A sample part)
■ Buy prototype parts bought or mock-ups built for packaging try-out

- Estimate product cost (request for quote or RFQ)
- Estimate production tooling and equipment cost
- Select design concept

The team will use the first level of parts to verify that the basic design meets or might meet the target requirements for the system. The A level hardware is used for:

- Product checking
- Cost estimating
- Time scheduling
- Manufacture, assembly, and test of prototype material

Revision Level B parts   The purpose of the B level of hardware is to verify the concept. This stage secures the parts used for build and prototype verification in the early phase of the project. Before the first release of the B level of hardware, the following actions should take place:

- Complete digital mock-ups
- Complete digital mock-ups for packaging study
- Do a preliminary verification of product capable of achieving the project targets
- Estimate part cost (RFQ)
- Estimate tooling cost and lead time (RFQ)
- Estimate manufacturing equipment requirements (RFQ)
- Select system engineering and supplier
- Identify any product structure to mating or interfaces to other vehicle components
- Complete early versions of the parts documentation

After the first release of the B level of hardware (a raw prototype part), the following actions should take place:

- Purchase prototype parts and tooling
- Verify feedback of parts and components by calculation and by bench testing

None of these parts is production representative with the exception of the dimensions. These parts are stereo-lithograph or other quickly constructed components that provide fit or dimensional data for the tooling of the component. If the tool is complex, the time spent at this level of hardware will be long to ensure the suitability of the hard tool. There is typically little-to-no software functionality at this level of prototype.

The B level component is typically used for

- Product checking
- Cost estimation
- Time scheduling
- Manufacturing and assembly and tests of prototype material
- Reservation of manufacturing capacity
- Any tooling design
- Procurement of tool

As with the A level parts, B level parts are targets for comment and preliminary decisions. The approach also applies to the B level service if we are following this model.

Revision Level C parts The main purpose of the C level of hardware is to secure the tooling requirements for the project. This phase releases approved parts and documentation for series production tooling and manufacturing equipment. Any hard tooling can be initiated after the start of this phase. It secures the manufacturing equipment required and the tools required for the build of pilot vehicles. This stage will be used to secure the initial sample approval (ISA) ordering.

Before entering the C stage, the following should be completed:

- Review manufacturing and assembly processes
- Test and verify manufacturing and assembly processes
- Include results of testing in update of the documentation revisions
- Tooling reviews with supplier and tool maker
- Estimate part cost (RFQ)
- Estimate tooling cost and lead time (RFQ)
- Estimate manufacturing equipment requirements (RFQ)
- Complete risk assessments before tooling commitment

Release of engineering documentation for tools with long lead-time and expensive tools should occur as soon as possible. Involved parties give approval to pass to stage C.

Before the first C release, the following documentation should be ready:

- Release all necessary engineering documentation for tooling and equipment
- Verify engineering solutions to fulfill the QFD (quality function deployment) targets
- Include in documentation tooling review comments

After the first C release, the following activities may cause changes that must be documented:

- Follow-up and comments with toolmaker
- Validation of parts of the production tools
- Assembly and test of product
- Inclusion of feedback results in updated documents
- Quotations from suppliers for
  - Product cost
  - Tooling cost
  - Lead-time
  - Manufacturing equipment cost
  - Manufacturing lead-time

This level of hardware should be representative of the production parts. They are typically built from some production tools, but were not put together on production equipment or using production processes. The C level is typically the level of component where the majority of the software development occurs, especially in cases of embedded development. A critical factor during the development of embedded software is the lack of hardware on which to code and test. In some enterprises, software teams will request printed circuit boards designed without regard to size so they can exercise the code as early as possible without having to worry about production-level sizing of the hardware. In embedded development, the wedding of hardware and software is more than intimate; this situation means the common software development tools are barely useful for development—the hardware is a necessity. The presence of hardware allows the use of in-circuit emulators, to plug into the target hardware (product) and perform the development.

The C level component is typically used for:

- Tool manufacture
- Ordering initial samples
- Procurement of production equipment

Revision Level P parts   The purpose of the P level (remember, these designations are arbitrary) hardware is to communicate to the organization that

- The part is approved from application and assembly in series production
- The part and engineering documentation reflect the production readiness of the product
- Validation has confirmed that part fulfills the QFD (if the team is using this tool)

Before entering the P stage, the following must be done:

- Complete all verification and validation
- Secure time plan for initial sample test and approval
- Approve product cost, tooling and equipment cost, and delivery schedule for production parts
- In the case of failing project targets QFD:
  - Assess risks
  - Develop an action plan
  - Propose solution for approval

The engineering documentation should:

- Include all the updates of the validation results
- Reflect the status of supplier tools and manufacturing equipment
- Have all involved parties get an approval (sign-off) to pass to P stage
- Have documentation of all product attributes
- Have bills of material ready for use by manufacturing, marketing, and after-market

After the first P release, the following activities may cause changes that the team should document:

- Manufacturing and assembly improvements
- Quality issues
- Administrative changes
- Initial sample approval status (disapproved)

3.3.3 *Production Issues*

Some of the following sections make sense only for the manufacturing enterprise; however, even the A, B, C, and P levels we discussed in the previous section can be generalized to embedded development and service design. We are putting explicit names on phases or stages of release. The approach reduces risk when planned well by the project manager and team.

3.3.3.1 *Bills of Material (Manufacturing)*

The manufacturing bill of material (BOM) is not constructed the same way as the engineering BOM. Engineers structure the manufacturing BOM to reflect the order of activities occurring during the fabrication of the product, rather than in the functional order most common to engineering BOMs.

Typically, the manufacturing BOM is located on the material requirements planning (MRP) or enterprise resource planning (ERP) system used by enterprise to manage production.

Manufacturing BOMs are usually tied to work sequences called "routings." The routings allow accounting departments to calculate cost at every work center.

Manufacturing BOMs can be modular; for example, they can be broken down in subassemblies, so that a subassembly that is common to several final assemblies needs to be specified just once. For planning purposes, a company may create a *phantom bill,* which allows acquisition of appropriate material when the true demand is not known.

The BOM applies to embedded software in the sense that the code and the integrated circuit comprise a subassembly.

A complex service may specify a bill of resources but does not usually require a BOM.

3.3.3.2 Process Flow Diagram

The process flow diagram uses the standard American Society of Mechanical Engineers (ASME) symbols for movement, inspection, storage, etc. and illustrates the fabrication activities that go into a product. The launch team identifies every production activity in the flow diagram. This document numbers each activity and these numbered activities will appear later in the process control plan (PCP) and the process failure mode and effects analysis (PFMEA).

Embedded software developers do not usually create a process flow diagram. Services, on the other hand, can use this tool to define the standard flow of service activities.

3.3.3.3 Special Process Characteristics

This concept is the same as special product characteristics, which is applied to processes to produce the design. These are important customer processes that must be included on the control plan per the ISO/TS 16949:2002 standard. A special process characteristic is one for which anticipated variation will often affect customer satisfaction with a product (other than S/C) such as its fit, function, mounting or appearance, or the ability to process or build the product. In other words, these characteristics must be under tight control per customer requirement. During the program definition phase, the launch team identifies the specific characteristics that will allow the product to meet the customer's expectations.

Special process characteristics can be defined for embedded development and service design. The special symbols and formal requirements provide a structure for recording and implementing customer desires.

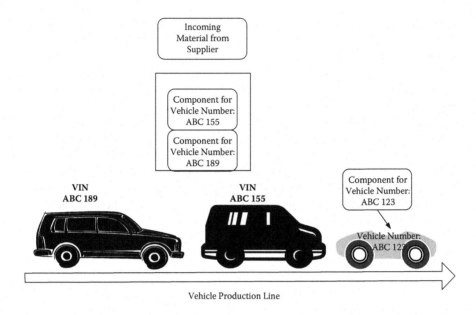

Figure 3.14   Line sequencing.

3.3.3.4   Line Sequencing

There are some vehicle manufacturers, typically automotive and not heavy vehicle, that require the incoming material to be synchronized with the vehicle production or vehicle identification number or some chassis identifier (see Figure 3.14). This feature is often tracked using bar codes, although with innovations and demands for radio frequency identification (RFID), this technique could soon be the preferred method for tracking the unique ID for each component. This technology places additional constraints and requirements on the product shipping and delivery systems. To be able to line sequence, the material or part ordering must be stable and predictable. Further, the level of customization of each unit must be minimal or the supplier of the part must be able to meet this variability in feature content for each vehicle. This is more of a concern for heavy vehicle production with often-erratic orders (also known as "drop-ins") and the large degree of variability in vehicle features in the vehicles using the component.

The line sequencing concept is relevant to embedded software also. It is conceivable that customers might desire customized software versions for their vehicles. The embedded software may identify itself to another node on the network, after which the master node will train the slave node; that is, the slave node and the master node have a common table of features so that all messaging and other system actions proceed harmoniously.

Line sequencing is also a component in industries that use an approach called "mass customization"; for example, some manufacturers in the clothing industry will have a garment sewn to the specific order of the customer, selling tailored clothing right off the factory floor.

The line sequencing idea also applies to services, particularly when the service provider has process detection mechanisms that allow the process to proceed down a different path. An example of this approach would be a provider of credit cards and other financial services such as Capital One.

Line sequencing and the related concept of attribute-based requirements (ABRs) require substantial support from information technology systems. Sometime these systems will have an explicit name like *configurator*.

3.3.4 Quality Issues

3.3.4.1 Quality Requirements

Thanks to the influence of the AIAG book on statistical product control (SPC) and others, quality requirements in automotive work are usually specified in terms of a process capability index called *Cpk*. The related index is *Cp* and can be calculated by dividing the tolerance interval by the measured six sigma variation of the product or process (three standard deviations from the mean on both sides) if and only if the process is under statistical control. Because *Cp* does not account for centering, the *Cpk* value provides a better idea of the quality situation. Engineers often abuse the *Cpk* index because they are calculating an alternative process capability index, *Ppk*, which does not have the statistical control stricture.

One of the issues with these capability indices is that they *must* follow a normal distribution. This requirement is not a problem when the value derives from the distribution of sample means because the central limit theorem says that the results tend toward a normal distribution no matter what the underlying distribution of the raw data. Unfortunately, we have seen production test equipment calibration data that did *not* follow a normal distribution (it followed a greatest extreme value distribution), yet the *Cpk* value appeared in the machine certification data.

Sometimes quality requirements are specified as parts per million (ppm). Although this value provides less information than *Cpk*, it has the advantage of being independent of the underlying distribution; that is, it is nonparametric. When establishing values of ppm, it is critical to use all the data; in short, if the ppm data are from end of line testing, the line operators and technicians should not be testing the component or product until they get a pass and ignore all previous failures of that part.

Cp and *Cpk* (*Pp* and *Ppk*) are meaningless in the embedded development world since we do not look for centering. The Rayleigh distribution (Weibull distribution with shape factor of two) provides some statistical information about the readiness for software product release.

Cp and *Cpk* may have some value if we are measuring deviations from expected duration of a service. We expect the user to wield these with caution.

3.3.4.2 Reliability Requirements

The concept of reliability ties together such concepts as quality, durability, and correct operating functionality and applies these as a function of time. Many automotive quality measurements (e.g., *Cpk*—an index of process capability) derive from as-tested quality, but supply no information regarding the behavior of the product over time. We suggest that reliability is a key measure of end customer satisfaction, particularly with work vehicles, where downtime means lost income.

The key to assessing the reliability of a new product (or an existing product) is to define the scope of desired results:

- Designed-in reliability
- Weakest point testing
- Predictive testing
- Predictive analysis (MIL-HDBK-217F)

Typical tools used during a concept or early-design phase are fault trees (rarely) or failure mode and effects analyses (design FMEAs or DFMEAs). The DFMEA looks at single modes of failure and endeavors to produce a high-quality qualitative assessment of points of weakness in the design. A DFMEA can be started as soon as the earliest design documents exist and it can be continued through the design process. DFMEAs are weak in that they do not consider multiple failure causes or cascading failure modes.

Weakest point testing can be accomplished with either multiple-environment overstressing or highly accelerated life tests (HALTs). These tests do not have predictive value, but they are powerful tools for discovering design weaknesses promptly.

Predictive testing generally falls into two camps:

1. Time-varying stress with Weibull analyses followed by an inference regarding product life
2. Accelerated life testing with an enormous number of samples (per Locks's Beta-related lower confidence interval calculation) and, usually, no failures.

The use of time-varying stress methods is less common than the reliability demonstration testing with no failures. The purpose of the strep-stress testing is to generate failures in order to create a Weibull plot that shows the failures at various stress levels. In general, the plot resembles a set of lines where each line represents a stress level. If all goes well, all the lines will show a similar shape factor, which then allows the reliability engineer to infer the probable life of the product under nominal conditions.

In the reliability demonstration mode, the product is subjected to stresses above nominal but well below the destruct level. This test mode produces a situation referred to as "accelerated testing." Although tables exist that allow prediction of confidence and reliability from *failed* parts, the more common approach assumes no failures will occur; hence, the reliability has been demonstrated. The weakness of this approach lies in the lack of failures—no one knows when the product will fail and, thus, no one has fully characterized the product.

Predictive analysis from the military handbook implies that the product has no parallel or redundant design blocks and that the behavior of the components is well understood. In our experience, the results from this kind of analysis yield a conservative assessment of the reliability. The downside of this is that the development team may overdesign the product in an attempt to increase the ostensive reliability.

We prefer the time-varying stress method preceded by multienvironment overstress testing or HALT. These methods should yield the most complete characterization of the product.

Embedded software reliability provides another class of difficulties. The goal is to produce software that has robust performance across the spectrum of input behaviors, be they normal or abnormal. Even if the fault-causing input behavior (stimulus) is unlikely to appear in the field, it will still have identified a weak spot in the software. The embedded system project manager must reject the expedient choice and ensure that the embedded developers correct the issue and improve the robustness of the software. Our experience suggests that test-bench anomalies are usually much worse and more common in the field.

Reliability in the service industries would relate to the quality of service during some quantum of time. To evaluate the quality of the service, a meaningful measurement system must be in operation. The auditors of such a system can represent the variation in quality of service with either a run chart or a control chart.

Environmental stress screening (burn-in)   Some vehicle manufacturers (and manufacturers in other industries) will require burn-in (see Figure 3.15) of the product during development of advanced prototypes and early production. Burn-in occurs when the supplier runs the part at some elevated temperature for a designated period of time in order to induce infant mortality in poorly manufactured parts. Burn-in is a special subtype of environmental stress screening. Vibration is another method to stimulate failure in cold solder joints (poor solder connection easily broken

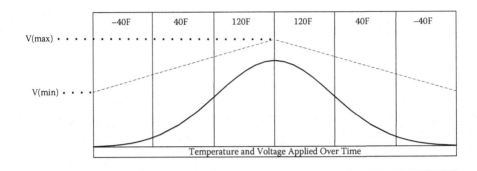

Figure 3.15   Burn-in test example.

by low amplitude vibration). An alternative to burn-in is the use of in-circuit testers to determine the status of surface mount and wave-soldered printed circuit boards.

Burn-in testing is probably the most common form of stress screening and stress auditing, wherein the units receive an elevated temperature for some quantum of time until marginal units fail (infant mortality). Burn-in typically occurs at the start of production when the manufacturing process is not under statistical control and substantial variation occurs. In general, manufacturing can reduce burn-in once the process stabilizes and the customer agrees. If the supplier wishes to promote the change, he can issue a supplier-requested engineering approval (SREA) request to the customer with data.

In some cases, stress screening/auditing may involve some amount of vibration testing, which is sometimes performed on electronic parts to elicit failures from weak solder joints.

Embedded development and service processes do not have a real analog with burn-in and environmental stress screening, although a synthetic load increase on a service can yield similar information about a process under stress.

Life expectancy   In our experience, vehicle manufacturers are under express life expectancy requirements. We suspect this situation to be the case in other industries also. The most common way to express reliability requirements goes like the following: "B10 life of 25,000 hours at 90 percent confidence." The B10 life occurs when 10 percent of the product fails in the field. Reliability demonstration testing provides an estimate of the confidence; the testing is usually a sequence of accelerated tests using Mitchell Locks's lower confidence limit based on a beta distribution calculation.

Unfortunately, specifying the B10 life and some hours is not sufficient to define the reliability of the product for the following reasons:

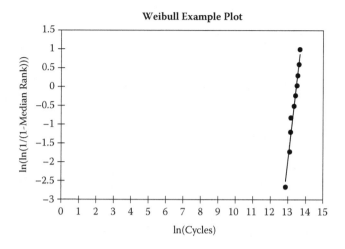

Weibull Example Plot

Figure 3.16　Example of a Weibull plot.

- We can plot the product data in a Weibull chart as censored units (stopped before failure).
- We can read the B10 life from the Weibull chart.
- The B10 life and the hours of usage do not define the slope of the line on the Weibull plot.
- The slope of the line on the Weibull plot describes the kind of failure mode (e.g., rapid wear-out accompanies higher slopes).

The slope of the Weibull plot (see Figure 3.16) is usually called β and is referred to as the *shape factor*. Designers should always require this shape factor in the reliability specification. The supplier can run samples to failure under accelerated testing in order to establish a realistic Weibull diagram for assessment of both the B10 life and β.

Embedded software reliability is less tractable to standard reliability analyses using Weibull plots and other hardware-oriented tools. However, it is possible for embedded software to degrade over time as nonvolatile storage suffers degradation [for example, Electrically Erasable Programmable Read-Only Memory (EEPROM) devices or flash memory] or if the random access memory fails. This type of degradation should not occur with production software. The most common method of eliminating confused code issues occurs through the use of so-called watchdog timers. A watchdog timer will force the software to reset—an observable or measurable event—if the code does not reset the timer.

Also, the defect arrival rate for embedded software can be modeled with a Rayleigh distribution, which is a Weibull distribution with a shape factor of two, which makes the probability density function resemble a log-normal distribution. In a log-normal distribution, the bulk of the data points are "lumped" earlier with a long tail to the right.

HALT/HASS Highly accelerated life testing and highly accelerated stress screening are advanced forms of testing used to determine the breaking point of the product. The intent of these tests is not to ensure that the product meets specification, but to determine where the product will be damaged—to find the weakest point or points.

HALT and HASS are part of a family of test-related activities. HALT is an acronym for highly accelerated life testing and HASS stands for highly accelerated stress screening. Synonyms for screening are 100 percent inspection or sorting. If using less than 100 percent inspection, the method is called HASA or highly accelerated stress auditing. All of the highly accelerated tests are qualitative tests—they have no predictive value.

The development or validation team can use HALT methods during development once the developer has parts that have come from production tooling. Frequently, less than five samples are used and once a failure mode occurs, the test facility can repair the part and put it back under test. This approach makes HALT testing cost-effective since the high sample count requirements of Locks's beta method for lower confidence limits are irrelevant to this kind of testing.

Any test that inflicts such a high level of stress that predictive methods are unavailable is a highly accelerated life test. Examples are the following:

■ Thermal shock (high speed cyclic temperature changes) label
■ Drop testing
■ High power spectrum density (PSD) vibration testing

In some cases, the designers or testers may wish to attack the sample in multiple dimensions. We can call this kind of testing HALT, but that is not the case—the testing can be done at a low enough intensity that life can be predicted from the Eyring equation. In all cases, we can call this kind of testing multienvironmental overstress testing. An example of this kind of testing would use vibration, humidity, and temperature in a special chamber. If we choose severe enough overstresses, the test becomes a HALT test.

In all cases of HASS and HASA, the team will incur important delays to the production cycle time. Additionally, they need to calculate the effect of the testing on the life of the product. Where HALT can save money during development, HASS and HASA always cost more money.

HALT work on embedded software occurs when, for example, the test group sends odometer messages at a much faster rate than normal, perhaps giving the illusion that the vehicle is moving faster than the speed of sound. Another approach might have the automated test sequence begin to send random odometer values to the software (examples are automotive here but can generalize to other products).

3.3.4.3   Product Assurance Plan

All of the activities that are performed to secure the quality of the deliverables for the particular phase compose the product assurance plan. The content of the product assurance plan has variation depending on the phase (for example, development phase activities would be different from verification phase activities) of the project. These actions are often tied to the gate targets, review points discussed earlier in section 1.3.6.7, or the description of the project phases.

- Plan and define
  - Understand/negotiate the project phase exit criteria
  - Review white books for similar projects
  - Review project with stakeholders
  - Review product requirements with stakeholders
  - Understand business objectives
- Product design and development
  - Understand/negotiate the project phase exit criteria
  - Design reviews
  - Software reviews
  - Perform DFMEA
  - Perform design for manufacturing and assembly
  - Reliability predictions
  - Aftermarket reviews
  - Key product characteristics identified
  - Systems test plan
  - Component test plan
- Process design and development
  - Key process characteristics identified
  - Production line design reviews
  - Process failure mode and effects analysis (PFMEA)
  - Control plan
- Product and process validation
  - Understand/negotiate the project phase exit criteria
  - Trial production runs
  - Run at rate
  - Process validation testing
  - Design validation testing
  - DFMEA follow-up
  - PFMEA follow-up
  - Systems and vehicle testing
  - Execute design verification plan and report (DVP&R)

3.3.4.4 Software Quality Assurance Plan

Institute of Electrical and Electronics Engineers (IEEE)-730 is a resource for setting the requirements for software quality handling. AIAG does not define software development processes and techniques to produce a quality product. However, the U.S. Department of Defense (DoD) has done so in several documents; for example, MIL-STD-498, Software Documentation and Development. The DoD documents metamorphosed into International Organization for Standardization (ISO) documents which, in turn, became IEEE documents; in particular, the IEEE 12207 sequence of documents.

1. Purpose
2. Reference documents
3. Management
    a. Organizational structure and controls of the software
    b. Software life cycle affected
    c. Task responsibilities
4. Documentation
    a. Software requirements specifications
    b. Software design description
    c. Software verification and validation plan
    d. Software verification and validation report
    e. User documentation
    f. Software configuration management plan
5. Standards practices conventions and metrics
6. Reviews and audits
    a. Software requirements review
    b. Software design review
    c. Preliminary design review
    d. Critical design review
    e. Software verification and validation plan review
    f. Functional audit
    g. Physical audit
    h. In-process audit
    i. Managerial reviews
    j. Software configuration management plan review
    k. Postmortem review
7. Test
8. Problem reporting and corrective actions
9. Tools, techniques, and methodologies
10. Code control
11. Media control
12. Supplier control
13. Records collection, maintenance, and retention

14. Training
15. Risk management

3.4 Cost

3.4.1 Request for Proposal

The customer initiates the request for proposal (RFP). It involves more than the desire for the development and material cost for the program, but rather a plan in addition to these. At this early state in the project, the RFP allows suppliers to submit proposals in response to the procuring organization's technical document, a product specification.

Typical contents of the response to the RFP are

1. Company historical information
2. Company financial information
3. Technical capability
4. Product technical information
5. Product availability
6. Estimated completion date
7. Estimated completion cost

The embedded development team will produce its share of the proposal along with the rest of the organization. Some managers will attempt to anticipate the complexity of the task, use a technique called "function point counting," and produce an estimate of hours needed to complete the job. The DoD often looks for a more sophisticated model called "COCOMO II."

The service industry can respond with an estimate of hours to deploy such items as peculiar support equipment (PSE) which is special purchase, one-off equipment. In the case of a service, this equipment might be more generic; that is, office equipment. In some cases, it may be point-of-sale devices.

3.4.2 Contract Types (fixed, etc.)

The contract type is important when the work is contracted and not internal to the procuring enterprise. The developing organization must deliver the product according to the contract. A firm fixed-price (FFP) contract can benefit the customer, leaving the developing organization to handle the impending delivery and any change management aggressively. Cost plus contracts allow the project conducting organization to adjust the contract according to the customer's changes or external demands.

Fixed-price incentive contracts provide a carrot to suppliers in an attempt to improve the requirements such as cost, schedule, or quality targets. The supplier once again takes the bulk, if not all, of the risk with a bonus when completing the objectives of the purchasing organization. The objectives can be material cost, delivery dates, or quality targets.

Cost contracts are used when there is considerable risk of the project being successful or when the supplier believes the risk is too high to take the project under fixed price alternatives. Software development can fall into this category. The customer often uses this method of payment when product specifications are incomplete or are ambiguous.

Cost plus incentive reduces the risk of the supplier while reimbursing the organization for costs that were taken to achieve the goals of the purchasing organization. However, this approach requires tight controls and an understanding of the suppliers' expenditures.

3.4.3   Pugh Matrix—Concept Selection

A tool for quantifying a number of design alternatives is the Pugh matrix. The example in Figure 3.17 illustrates the use. This tool uses a number of attributes that are key to the success of the design (leftmost column). The tool user prioritizes these attributes and compares each of the possible concepts. The design team evaluates the proposed design solutions by giving a value to the achievement of the attribute. This comparison can also include any present solution as a baseline. This information is collected in an attempt to quantify the competing designs and select the most appropriate solution. The design generating the highest numeric value exhibits a greater quantity (subjective) of the evaluating criterion and is the top design candidate.

3.4.4   Crashing Program Definition

When crashing the project, the project manager attempts to use cost against schedule as a trade-off to compress the schedule the greatest amount while slightly increasing the cost.

Fast tracking a project means staging phases or subphases as quasi-parallel activities rather than the more common sequential tasking. This situation often results in rework and always results in increased risk.

3.4.5   Program Definition Risk

3.4.5.1   Funding Risk

Insufficient funding of both capital and expenses is a major risk along with inadequate early funding for engineering development and testing and inadequate early production financial support. Long lead-time items

Pugh Matrix		Reference Design		Concept 1		Concept 2		2007-09-06 Concept 3	
Evaluation criteria	Priority—5 is high	Rating	Weighted	Rating	Weighted	Rating	Weighted	Rating	Weighted
Quality									
Maintenance support	5	0	0	−1	−5	−1	−5	3	15
Quality and Reliability	5	0	0	0	0	1	5	2	10
Complexity–user friendly	4	0	0	−2	−8	2	8	1	4
Assembly process	5	0	0	−2	−10	1	5	1	5
End of Line diagnostics and troubleshooting	5	0	0	1	4	1	5	1	5
Parts availability	5	0	0	−1	−5	−1	−5	−3	−15
Cost									
Product cost	4	0	0	−2	−8	1	4	3	12
Development cost	3	0	0	−1	−3	2	6	3	9
Assembly time	5	0	0	−2	−10	3	15	2	10
Maintenance	3	0	0	−3	−9	1	3	1	3
Repair time	5	0	0	−1	−5	−3	−15	−1	−5
Training/Education	5	0	0	−1	−5	−1	−5	4	20
Feature									
Product expandability	3	0	0	−1	−3	−1	−3	2	6
Scale–ability	5	0	0	−1	−5	−1	−5	1	5
Feature flexibility	5	0	0	0	0	0	0	1	5
Configurability for second life application	4	0	0	−1	−4	−1	−4	−2	−8
Special tools	5	0	0	−2	−10	−1	−5	4	20
Field replaceable	5	0	0	−1	−5	1	5	2	10
Backwards compatibility	3	0	0	−1	−3	0	0	1	3
Delivery									
Development speed and flexibility	5	0	0	0	0	4	20	3	15
Total Score			0		−94		29		129

Leader:
Engineering:
Aftermarket:
Product Planning:
Product Planning:
Marketing:
Eng. PM:

Figure 3.17   Example of a Pugh matrix.

are always a critical issue, especially when Materials Resource Planning (MRP) rules do not allow for material acquisition using a partial bill of materials. Risk increases when development commences without consideration for production issues. The development decision involves commitment to production which, in turn, must be supported by funding.

3.4.5.2   Reducing The Risk

To reduce the risks to the program definition, the project team must:

- Calculate funding thoroughly.
- Consider programs started on short notice with a jaundiced eye.
- Fund abundantly for production preparation. A significant initial subset of this profile is the engineering development funding spent on production preparations. If this funding profile is changed, the effect on transition needs assessment.

The team should request early commitment of production funds—while development is still ongoing—for

- Tooling,
- Long lead materials,
- Production line startup.

The fly-before-buy style of development and product launch tends to drive into the too-late category. Extreme concurrency can cause imprudent commitments. For all programs, the goal is an optimum that results in low engineering development risk and controlled transition to production. Early availability of enough funding from engineering development and purchasing financing is indispensable to an efficient handoff from engineering to production operations.

3.4.5.3   Technical Risk Assessment

In our experience, if any area needs major attention for both customer and supplier, it is *technical risk assessment*. Technical risk assessment can determine if a program is even feasible and it also provides for tracking of various factors during the course of the program.

3.4.5.4   Reducing The Risk

Risks are better addressed sooner than later. To reduce the project risk exposure from a technical perspective, the team can:

- Require technical risk management in specification.
- Identify all areas of risk as early as possible in the development cycle. Determine a specific set of tracking indicators for each major

technical element (design, test, and production) and for cost and management.

■ Develop plans to track, measure, assess, and adjust for identified risks using a disciplined system that is applicable by managers from a variety of positions within the customer and the supplier organizations. This system provides a continuous assessment of program health against quantifiable parameters.

■ Understand risk drivers using qualified design and production engineers to identify and reduce program technical risks.

■ Highlight technical problems before they become critical.

■ Avoid hasty shortcuts, review operational profiles, and use existing analysis tools while implementing the technical risk assessment system.

■ Structure test programs to verify and resolve high-risk design areas.

A technical risk assessment system should provide all levels of management with (1) a disciplined system for early identification of technical uncertainties, (2) a tool for instantaneous assessment of current program status, and (3) early key indicators of potential success or failure. To be effective, project management should initiate a technical risk assessment system at the start of the program to function during the development and production phases.

3.4.5.5   Schedule Risk

Inadequate resource availability and management pressure to reduce the expenditure for the project gives the illusion that managers are good stewards of the organization. The project manager must report to management all estimates of durations and schedule needs and defend expenditures and resource requirements.

3.4.5.6   Reducing the Risk

Do the following as needed:

■ Understand all deliverables and scope of the project
■ Create a detailed WBS
■ Use historical data to produce duration estimates where possible
■ Rely on experts to produce estimates for their portion of the WBS that are not *point sources* but have a range of possibilities and probabilities
■ Spend time sequencing the WBS activities to keep minimizing critical path tasks
■ Understand resource characteristics (both human resources and equipment) availability

3.4.6 Management Support

The project team should involve the appropriate management in the updates of the project status. The recipients of this update should be identified within the communications plan and the frequency, type, and format of the information.

> The team should update management at the conclusion of every product quality planning phase to maintain their interest, plus reinforce their commitment and support - AIAG[2]

Management support of the project is essential. In this phase, the program definition, the support can be generated around the business case (ROI or IRR) or the competitive advantage obtained from project (first to market or meeting a feature of the competitors). No matter the phase, management understanding of the state of the project and the reason for undertaking the project should be reinforced. Often, the team will not seek management support until the project is in crisis.

3.5 War Story

3.5.1 Cutting Edge Components

What follows is a technical risk war story. An instrument cluster project is in process. The team determined a certain amount of on-board memory was required. The current available production micro-controller did not have enough of the required flash memory and the supplier was working on a pin-for-pin replacement part that had additional flash and RAM memory. The development proceeded with the current production controller, since the controller with the additional memory would be available toward the end of the development process. This decision allowed the project to proceed without having an effect on the required introduction date. The new micro-controller would be available as the software developers coded the last features and the additional memory would be available for these final features. However, in the course of the micro-controller development project, the supplier determined the project was not a prudent business decision due to technical challenges. The development on the advanced micro-controller was terminated and the cluster project remained with an inadequate micro-controller. The developers retooled the design and added external memory to the electronic control unit (ECU). This chronology created a delay in the project and presented the organization with cost increases that were unplanned.

3.5.2   Herding Cats

Multiple divisions of a company accepted a project. At first, the goals of the participants seemed to be common enough to have the divisions collaborate to produce the product. It soon became apparent that it was going to be difficult to meet all the stakeholder's goals in one common project. However, the project continued to move forward with the expectation that all stakeholder targets would occur. In short, once the team determines the project will not meet the fundamental goals, it is time to use the kill-point concept and put the project out of its misery.

Chapter Notes

[1] Automotive Industry Action Group, *Advanced Product Quality Planning and Control plan* (APQP), (Southfield, MI, AIAG 1995) p. 7.

[2] Advanced Product Quality Planning and Control Plan (APQP), (Southfield, MI, AIAG 1995) p. 11.

Chapter 4

Product Development

4.1 Product Development Overview

The product development team (example in Figure 4.1) consists of various engineering disciplines needed to achieve development and implementation of the product. During the product (service, hardware, or software) development phase, resource availability is often the most critical issue. Not shown, but included, is the customer. If a supplier is developing a specific product for a particular customer, the customer may well become involved in the development. They may not have deliverables to the project during this phase; however, customers will often participate in the project—a situation detrimental to efficient progress due to customer interference.

4.1.1 Phase Objectives

This section is where the developmental work occurs. The objective is to design the product to meet stakeholder expectations identified and specified—earlier with the customer being the most significant stakeholder (see Figure 4.2). Be it a product, embedded software, or a service, the design occurs during the development phase. In the case of a physical product, the concept of design for "X" (DFX) becomes important (the "X" refers to reliability, manufacturing, testing, quality, or cost).

When it comes to the actual design work, reuse of previous work has several benefits (applies to hardware, software, and services):

1. It shortens lead times (expedites delivery) because assemblies, code, or procedures already exist and do not require extensive redesign work.

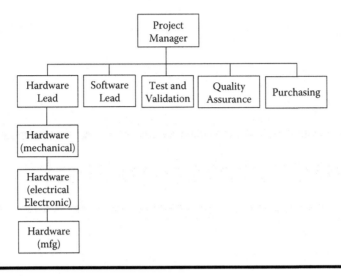

Figure 4.1   Example of a product development team.

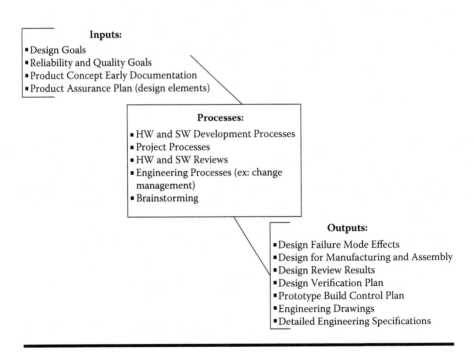

Figure 4.2   Product development interaction.

2. It saves on material cost due to volume increase (applies solely to hardware).
3. It provides consistency for manufacturing process consistency (low product variability and support) including software and services.
4. It provides a risk mitigation history for the subassemblies, code, or service, including field data.

It is possible to develop modular hardware designs which allow for reusability. This modular approach, coupled with some available part numbers and descriptions for the module, will allow the hardware and software engineers to use modular parts repeatedly.

Production processes are reusable too, minimizing the need for creation of new line equipment. Even if it is necessary to develop a new piece of hardware, the historical product development information helps to mitigate design risk. Service processes usually do not have the capital risk, but these processes also benefit from modular design.

The supplier works to develop the product through a number of iterations of hardware or software (see hardware and software discussion, Chapter 1). The developing organization may perform engineering verification on proposed aspects of the design solution.

4.1.2 Inception

The inception phase can occur during the voice-of-customer project inception phase or can be a unique start for this phase. If this is a separate inception phase, the discussion relates only to the development of the product. For our purposes, we will treat this process as starting the design activities. A kickoff event is not out of place when commencing this phase.

4.1.3 Planning

The project manager and the launch team develop a plan to support development activities and address those risks that are intrinsic to the product development cycle. The planning phase considers how best to achieve the objective of designing the product: resources, time and material, and cost.

Planning includes *securing* the design work through quality assurance activities such as design reviews, code reviews, and other controls. A requirement of this phase may also be to provide support for after-market documentation requirements. As always, if it is on the scope list, there must be a plan.

4.1.4 Implementation

During the product development phase, implementation lies in choosing a particular design concept and developing the product. The engineers create detailed specifications (engineering specifications) and hardware and software designs. Specifications from the earlier phase serve as inputs. This phase can include some testing or verification work on the concepts, but the bulk of the work concerns the creation of the product and the quality assurance activities.

According to AIAG, the outputs for this phase are[1]

- Design failure mode and effects analysis (DFMEA)
- Design for manufacturing and assembly
- Design verification
- Design reviews
- Prototype build control plan
- Engineering drawings (including Math Data)
- Engineering specifications

If the development team is running a project to deliver a service, then many of the product-oriented documents and controls do not apply. However, in the services industry, the *service* is the product. An enterprise can design a service, verify it, run prototype activities, and produce a specification—even drawings if they document the flow.

4.1.5 Regulation

Project reviews and procedures such as Design for Manufacture and Assembly (DFMA) and DFMEA provide feedback on whether the efforts undertaken are on track to produce the product. To ensure quality of the product, the team will use systems and software reviews. Component testing and engineering tests drive design and development. This feedback results in corrective actions, other levels of supplier selection, and modification of the design concept. The project manager and the launch team monitor the project to determine if they are meeting the goals of this phase, with corrective actions initiated to formalize action against anomalies. Tools to facilitate control are action item lists and identified metrics for the project and associated engineering activities.

4.1.6 Termination

The phase termination process provides a review of the deliverables determined during inception and ensured while planning and built during implementation with feedback from the controlling activity. The termination process can have contract reviews and supplier delivery audits and reviews of the things delivered in the implementation portion of this phase. We might ask ourselves some questions, such as the following:

- Were the results of the DFMA activities acceptable?
- Did the design reviews occur? Were they of suitable quality? Are they documented sufficiently?
- Are the drawings available? Did we release them?
- Did the team deliver an acceptable control plan?

4.2 Delivery

4.2.1 Crashing Product Development

4.2.1.1 Schedule Compression

If our in-progress delivery dates slip or customers pull in their need for the product, process, or service, the project manager will compress the schedule (also known as "crashing"). This situation can happen often and the project manager's response to this demand is critical to project and organizational success. Compressing the schedule increases risks. Sometimes the payoff is greater than the risk effect. Sometimes, other actions occur to mitigate the increasing risk due to the schedule compression.

The following formula provides a simple approach to coming up with a value for the compression:

$$1 - \left[\frac{Calendar\ time\ schedule}{Nominal\ expected\ time} \right] \times 100 = Schedule\ Compression$$

Attempts to compress a schedule to less than 80 percent are not usually successful.[2]

4.2.1.2 Crashing Limitation

Sometimes referred to as "variable factor proportions," the law of diminishing returns says while equal quantities of one factor increase and other factor inputs remain constant, all things being equal, a juncture occurs beyond which the increase of another unit of the variable element results in diminishing rates of return.

Example: Adding laborers to harvest a wheat field may increase productivity. However, continuing to add laborers will, at some point, create a condition where each additional laborer will not perform the same amount of work because he has less and less of the fixed amount of land to work. This insight is applicable to all enterprises. The threshold at which adding resources has a negative effect is hard to identify and varies with technical constraints such as production techniques. Also, if we built a map of the laborer inter-relationships, we would see communication becoming more complex.

4.2.1.3 Crashing Risk

It can become obvious that the execution of a project will not meet the required due dates. One tactic used to enhance the probability of project success is to put many resources into the project. When all or many of the resource constraints are abandoned, the cost for a project becomes enormous. Taking this action all but guarantees the project will not happen within the original accepted budget. In short, the project manager and team should avoid crashing whenever possible.

4.2.1.4 Fast Tracking

When the team compresses the schedule by overlapping activities that are otherwise serial activities, we call it "fast tracking" (see Figure 4.3). If the team delivers the project using serial phasing, then each phase comes

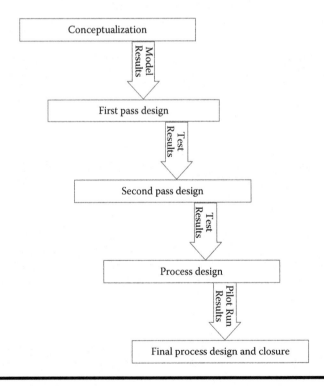

Figure 4.3 Fast tracking.

to closure before the next phase begins. However, when trying to expedite a project, the team performs some of the execution in parallel. For example, if the project arrives at the closing phase of detailed development or specification-writing, the team might choose to have the development phases overlap and begin the actual design work before having all of the specifications in hand.

4.2.1.5   Fast Tracking Risk

When fast tracking, there is an increase in project coordination risk because the sequential actions needing output from a previous phase occur concurrently. In the example in the previous paragraph, design development starts before securing the design documentation. Starting production work before completing the design is another example of fast tracking.

4.3   Product Integrity and Reliability

4.3.1   Design Failure Mode and Effects Analysis

The design failure mode and effects analysis (DFMEA) is a qualitative tool that allows design engineers (and other team members) to anticipate failures, causes, and effects in a product. The DFMEA is different from the process failure mode and effects analysis (PFMEA), which the manufacturing engineers use to anticipate failures in the production process. Additionally, the AIAG also specifies the requirements for a third form of FMEA called a machinery failure mode and effects analysis (MFMEA). The maintenance staff in a manufacturing organization uses the MFMEA to anticipate and prevent failures in production equipment.

When creating a DFMEA, the engineers must consider the failure modes to be observable *behaviors* on the part of the system, subsystem, component, or subcomponent. In some cases, the behavior may be observable using a special tool such as an oscilloscope. Nonetheless, the *failure mode* is an observable behavior or response. The cause of a failure mode is a *stimulus* known to produce such a response. When the FMEA team approaches the creation of the document using this method, they can expect a high level of document consistency and logic.

The AIAG also specifies something called an *effect*. An effect is whatever the team defines it to be, although a frequent choice in the automotive world is the *effect* on the driver/operator of the vehicle. For example, a malfunctioning speedometer would be a violation of regulatory requirements which would force the driver to have to manage the issue, perhaps by taking the vehicle to a dealer or other repair facility.

A typical generic collection of failure types is as follows:

- Complete failure,
- Partial failure,
- Aperiodic failure,
- Periodic failure,
- Failure over time,
- Too much of something,
- Too little of something,
- A value never changes (frozen).

All FMEAs try to provide a numeric assessment of three different categories of consideration:

- Severity,
- Occurrence,
- Detection.

Severity receives a numeric value from a ten-point scale (1–10) with 10 being the highest value, which involves safety and regulatory requirements. Analogous to severity, occurrence and detection both use a ten-point scale. In each case, "1" is the lowest value and "10" requires action. From the three values, we calculate a risk priority number (RPN) as follows:

$$severity \times occurrence \times detection = RPN$$

Once we calculate the RPN, the form provides for recommended actions, assigned responsible parties, and due dates. Often, the team will sort the form by RPN with the highest values at the top. The team may decide to manage only the situations where the RPN is above some agreed on value. The team can review items with a high severity regardless of the values for occurrence and detection. This approach will pick up lawsuit situations where the severity is high and occurrence is low and detection is easy but late.

Once the team has managed the recommended actions, the FMEA team should recalculate the RPN—the form provides special columns for that purpose. If the recommended actions accomplish the goal, the RPN will drop below the threshold value for consideration as an issue.

An alternative to the FMEA approach is the *fault tree*, a technique favored in some government acquisitions and in the nuclear industry. The fault tree has the advantage accounting for multiply triggered events, whereas the FMEA examines single points of failure. The fault tree is a visual analog to Boolean notation. The downside of the fault tree is that it requires special software to support its complexity and to perform *grammar* checks on the logic. Fault trees are also labor intensive.

Both fault trees and FMEAs are *qualitative* approaches to managing and anticipating failures in the system. They are only as good as the effort put into them and the benefit of FMEAs in particular does not appear to have been well quantified by research.

Note also that the FMEA is a document required for production part approval process (PPAP) and is part of the reliability engineering armamentarium.

4.3.2 Design for Manufacture and Assembly

Designing for manufacturing and assembly (DFMA) does not mean waiting until the manufacturing stage before designing. To achieve the best return on the time invested in designing for efficient production, it is necessary to start early to reap the most benefit (see Figure 4.4). This requires that the launch team work the production and assembly requirements simultaneously with the development project or, at the least, the design staff has some idea of the present state and future trends of the production floor. *This includes any external suppliers in the development chain for the product.* In general, the factors that influence the successful DFMA are

- Product complexity
- Staff experience and knowledge
- Organizations available engineering and manufacturing tools
- Available time to implement
- Product cost targets
- Product competitive environment

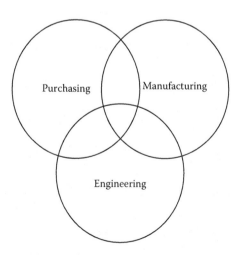

Figure 4.4   Design for manufacturing and assembly.

The design complexity has a significant effect on the cost. If the product has many pieces requiring manual assembly, the cost of goods sold increases. A simple way to understand the cost of complexity is to multiply the yield at multiple points in a process until we achieve a result called "rolling throughput yield." The approach is analogous to calculating serial reliability. What we learn is that additional operations reduce yield—the cost of complexity.

DFMA is an attempt to solve problems before they happen, anticipating and acting instead of complaining after the process is already in place. Used effectively, it saves time by eliminating process design rework and it improves the process yield.

Note that the DFMA approach can generalize to services and embedded development as well. A complex embedded project may have a communications problem among software modules that is solvable by reducing modules and following structured approaches (structured software or object-oriented development).

4.3.3 Prototype Builds

During the course of the project, the developer can use prototype parts for analysis of fit as well as some small analyses of function. The choice of approach depends on the stage of the development project and the end use of the components. Early prototype parts are used for fitment testing (making sure components fit together) and limited material and performance tests. The testing must be within the constraints of the material; for example, performing a vibration test on a stereolithograph part is almost certain to fail.

Figure 4.5 shows one sequence using prototype parts. A working model may not resemble the final part in configuration at all, but it functions sufficiently to elicit a critique from the customer. In some cases, the prototype will closely resemble the final part, but is still nonfunctional. The device that creates parts with stereolithography is called a stereolithography apparatus (SLA). Frequently, the parts themselves are called SLAs after the name of the machine. The SLA creates the part by depositing a resin in layers to build a representation of the product. The SLA part has no mechanical strength. Hence, the SLA is a nonworking representation of the form and fit of the final product.

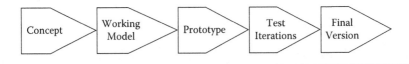

Figure 4.5   Prototype parts.

In order to use the prototype parts, the development team requires an understanding of the limitations of the parts and the manufacturing processes that generated those parts. Therefore, the development team should develop a prototype control plan. The plan is usable for prototype parts and pre-production parts. Unique part numbers or identifiers help identify and track parts.

4.3.4   Key Product Characteristics

The definition of key product characteristics is one of the gifts of automotive manufacturing to all other kinds of production. It is quite impossible to cost-effectively measure every possible characteristic of a given product. However, it is possible to define the most significant characteristics as key product characteristics. For example, the front of an instrumentation cluster may have significant appearance requirements, but it is usually not necessary that the back of the product (invisible to the operator) have the same level of appearance quality. Hence, the front appearance and its definition is a key product characteristic. A key product characteristic (KPC), Figure 4.6, is a feature of a material, process, or part (including assemblies) where the variation within the specified tolerance has a significant effect on product fit, performance, service life, or manufacturability. The same idea applies to services where the measurements in the back-end of the process are not relevant to the customer who only sees the front-end of the process.

It is not the special symbols that are so important but, rather, the idea of *key* product characteristics. An example from a mechanical point-of-view would be any location where one part fits with another part. The figure shows symbols that automotive companies use in the United States.

We have seen instances where a failure to systematically define customer requirements led to material choices that could not meet customer needs.

Customer	Term	Symbol
Chrysler	Safety Item	<S>
Ford	Critical Characteristic	▽
General Motors	Safety Compliance KPC	▽
Chrysler	Critical Characteristic	◇
Ford	Significant Characteristic	◇
General Motors	Fit/Function Key Product Characteristic	◇

Figure 4.6   Key product characteristics.

For example, in one case the customer had clearly specified that the quiescent current draw for the product should not exceed one milliamp, but the design engineer selected communications components in the hundreds of milliamps thereby producing a battery drain issue. All of this uproar could have been avoided by researching the customer specification and defining key product characteristics up front in the design process. In this particular incident, the choice delayed the launch of the product and a redesign and revalidation of the product occurred at a cost of tens of thousands of dollars in wasted effort and time.

4.3.5 Design Verification

During the design and development cycle, there are tests conducted to ensure the probability of the design to meet the final requirement. This testing characterizes physical traits such as thermal characteristics or other specific electronics requirements. These tests can and are often part of the development effort. When part of the development process, we call them "engineering verification tests." These tests reduce the risk at the design verification phase of the project by verifying performance during the development effort, allowing the team to confirm or refute models and allow for redesign.

4.3.6 Product Development Risk

4.3.6.1 Design Requirements Risk

Design requirements derive from operational requirements and often evolve during the progress of the design. Intangible and improperly derived design requirements are a cause of high risk.

4.3.6.2 Reducing the Risk

To reduce the design risks in the product development

- Develop design requirements in parallel with the product concept (sometimes called a "product description"—a high level document). Define them in the requests for quote such that the customer can evaluate different suppliers comparatively.
- Define design requirements in terms of measurable performance.
- Ensure that approval applies to reliability, maintainability, FMVSS safety requirements, corrosion prevention, parts standardization, and all related systems engineer activities.
- Make sure that suppliers are responsible for ensuring that subtiers have complete and coherent design requirements that flow down customer requirements including measurable parameters and task performance.

4.3.6.3   Design Policy Risk

The execution of systems disciplines in order to reduce risk is the job of suppliers. The presence or deficiency of documented corporate policies, backed up by a frequently audited quality manual (automotive uses ISO/TS 16949 as the fundamental quality standard-derived from ISO 9001:2000), has a direct influence on the degree of product risk associated with new product development. It is not so much the documents themselves that make a low-risk company as the discipline exhibited by the company that maintains and executes quality policies and procedures.

4.3.6.4   Reducing the Risk

To mitigate the policy risks the organization must assess the level of risk acceptable and develop systems to support. This is done by:

■ A quality system is in place.
■ Policies and practices contain—implicitly and explicitly—lessons learned from previous development efforts.
■ Audit data are available to substantiate compliance to proper engineering practices.
■ Policies and procedures support design for maintenance, assembly, manufacturing, and testing.
■ Engineering design has the documented responsibility not only for development of a low-risk design, but also for specification of test requirements and design for production and support.
■ Documented design review expectations are easily accessible.
■ Design emphasis relies on execution of design fundamentals, disciplines, and procedures known to lower design risk.

4.3.6.5   Design Process Risk

The design process must exhibit sound design policies and procedures and best-practice engineering discipline by integrating factors that shape the production and service of a product through its life cycle. Frequently, suppliers and customers will discuss concepts with little thought for the feasibility of actually making a product (for example, choosing an electroluminescent display when the price is exorbitant). This omission is then carried forward into design, with voids appearing in manufacturing technology and absence of proven manufacturing methods and processes to produce the system within cost targets. One of the most common sources of risk in the transition from development to production is failure to design for production. Some design engineers do not consider in their design the limitations in manufacturing personnel and processes. The predictable result is that an apparently successful design, assembled by engineers and highly skilled model shop technicians, goes to pieces in the factory environment when subjected to full-rate production. A design should not survive the review process if it cannot survive full-rate production without degradation.

4.3.6.6 Reducing the Risk

To mitigate the design process risks to development, the organization must "know" good development processes. They must be keenly aware of their development processes and:

- Explore the potential to produce a reliable, high-quality system during the verification and validation phases (separate items in the automotive universe) using producibility analyses.
- Manage missing information in new production technology projects (for example, some selective solder robots) and manufacturing methods particular to the design of the system, subsystems, and components during engineering development. The standard method of manufacturing qualification in the automotive industry includes pilot runs, PPAP runs, and run-at-rate studies.
- Shun design dependence on a single unproven manufacturing technology (the *silver bullet*) for critical to quality performance parameters.
- Integrate producibility engineering and planning as an integral element of the design process. Establish close coordination between production and design engineering from the start. Foster integration of life cycle factors in the design by forming design teams with production engineering and support area representatives. Manufacturing coordination is part of production drawing release. Production engineers participate in design concept development and design engineers participate in production planning to ensure design compatibility with production.
- Ensure the design answers both performance and producibility considerations for product and component packaging.
- Evaluate the design to ensure that manufacturability and supportability factors are being incorporated. Manufacturability and supportability design changes are combined as early as possible to reduce cost. Verify these changes through testing.
- Support cross-training of engineers in design and manufacturing disciplines actively. Design engineers stay abreast of developments in manufacturing technology that would affect the design. In practice, this item rarely occurs.

The design process describes all the actions that result in a set of drawings or a database from which the team constructs a model for verification of specification compliance. They develop design criteria and check them before the final system design. Production design occurs in parallel with the other elements of the design process.

4.3.6.7   Design Analysis

Engineering design involves many specialized analyses, most of which point toward meeting desired performance specifications.

4.3.6.8   Reducing the Risk

To reduce the design risk, the project can use engineering techniques such as:

- Perform stress and stress/strength analyses to ensure that applied values of all parameters specified in the derating, margin of safety, and safety factor criteria for all component parts and materials meet those criteria.
- Perform worst case tolerance analyses to verify that the system design performance remains within specified limits for any combination of component part parameters within the limits of their own allowable tolerances. The engineers can use root sum of the squares or Monte Carlo methods.
- Perform sneak circuit analyses to detect such unexpected failure modes as latent circuit paths, timing errors, or obscure cause and effect relations that may trigger unintended actions or block desired ones without any part failures occurring.
- Perform FMEA in order to understand the effect of each component part failure on overall design performance and system and equipment supportability. Analyze each component part for the purpose of reducing these effects to a minimum through design changes.
- Conduct a thermal survey on electronic systems.
- Apply other analyses, such as fault trees, mass properties, system safety, maintainability, life cycle costing, fault isolation, redundancy management, and vibration surveys.
- Use the results of these analyses to revise the design to reduce design risk and update the analyses by changing the design.

The team develops design analysis policies before the final system design, but they should update and refine them as they gain experience during development. Their use is complete, except for engineering changes to correct failures, at the conclusion of the design process.

4.3.6.9   Parts and Materials Selection

Low-risk designs allow parts and materials to operate well below their maximum allowable stress levels. Performance-oriented military programs often attempt to use these same parts and materials at much higher stress levels. Pursuit of interoperability and parts standardization also may introduce similar risks. These choices often are made by using mathematical models and generic handbook data that are imprecise. The resultant high risk may not appear except by testing—which is too late to avoid extensive corrective action.

4.3.6.10 Reducing the Risk

To reduce the risk due to material in the design, the project must consider:

- Use design criteria for part operating temperatures (except semiconductors and integrated circuits). These criteria apply to case and hot-spot temperatures. Of all the forms of stress to which electronic parts are susceptible, thermal stress is the most common source of failures. The thermal stress guidelines that are highlighted have been instrumental in reducing the failure rate of electronic equipment by up to a factor of 10 over traditional handbook design criteria.
- Lower junction temperatures of semiconductors because failure rates of semiconductors decrease a lot.
- Determine the limiting values of operating temperatures for all electronic parts in a design both by analysis and by measurement. In many cases, these temperatures are determined by thermocouples or pyrometers.
- Include customers and suppliers as participants in design policies and in parts and materials derating criteria for all parts used in their products, specifying design limits on all items for which reliability is sensitive. Derating is a technique for reducing risk and ensuring parts/components are never pushed beyond their design limits.
- Constrain designers with preferred parts lists. Designers must use the selected standard parts when they meet system requirements or justify nonstandard parts.
- Use engineering development testing as a minimal set of tests to qualify the proposed design. This testing ensures the product development team is on the right track with respect to the design and the component selection. Additionally, this confirms the supplier can meet the design challenges.

Parts and materials selection and stress derating policies should be initiated at the start of hardware development. Supplier design reviews are the primary mechanism to ensure compliance with these policies.

Note that materials selection is critical to embedded development. The software development team should participate in micro-controller choices and sourcing to avoid unnecessary controller changes.

4.3.6.11   Software Design

Many vehicles worldwide now depend on software for operations and maintenance. In most cases, the system cannot even be tested without the product software. Software defects can cause vehicle failures. It is essential to allocate system requirements between hardware and software such that failure to meet requirements can be identified.

4.3.6.12   Reducing the Risk

Software design risks can be reduced by:

- Functional requirements are allocated either to hardware or software, as appropriate, at design start.
- Best practices with respect to design policies, processes, and analyses are employed, including but not limited to the following:
  - Rigorous configuration control.
  - Design teams and modular construction.
  - Structured programming and top-down design. Note that this applies to object-oriented software also.
  - Structured walkthroughs and/or inspections. Michael Fagan demonstrated in the 1970s that code inspection occurs earlier than testing and usually elicits *different* errors than those found by testing.
  - High-quality documentation.
  - Traceability of all design and programming steps back to top level requirements.
  - Reviews of requirements analyses and design processes.
  - Thorough test plans developed and used from design start. Test documents also need requirements flowdown.
  - Compliance with standards.
- Computer software developers are accountable for their work quality, and are subject to both incentives and penalties during all phases of the system life cycle.
- A uniform computer software error data collection and analysis capability is established to provide insights into reliability problems, leading to clear definitions and measures of computer software reliability.
- A software simulator is developed and maintained to test and maintain software before, during, and after field testing.
- Security requirements are considered during the software design process.

Software design practices should follow a disciplined process analogous to proven hardware design practices. Design schedules for software are often concurrent with the hardware schedule making the testing situation even worse.

4.3.6.13 Design Reviews

Well-constructed and recurring reviews can have a positive influence on the design outcome. When many eyes and brains review a project and the design details, the likelihood is that design errors reveal themselves. When a group of engineers with varied experience meet and critique the design, they uncover software or hardware problems of design. When the team practices design reviews rigorously, it can eliminate or reduce expensive testing by removing problems promptly.

While customer specifications usually require formal design reviews, they often lack specific direction and discipline in the design review requirement, resulting in an unstructured review process that fails to fulfill either of the following two main purposes of design review:

1. To bring to bear additional knowledge to the design process to augment the basic program design and analytical activity;
2. To challenge the satisfactory accomplishment of specified design and analytical tasks needed for approval to proceed with the next step in the material acquisition process.

4.3.6.14 Reducing the Risk

We can reduce some of the risk inherent in design reviews with the following ideas:

- Develop a design review plan and solicit approval from the customer.
- Flow down design review requirements to supplier tiers to ensure good internal design review practices and to provide timely opportunities for both the supplier and the customer to challenge the design at various tiers.
- Select customer and supplier design review participants from outside the program under review on the basis of experience and expertise in challenging the design.
- Solicit representation from manufacturing, product assurance, and purchasing functions who have authority equal to engineering in challenging design maturity.
- Use computer-aided design analyses whenever available and include reviews of production tooling required at the specific program milestone. Some examples of computer-aided design analyses are:
  - Finite element analysis for temperature and stress/strain,
  - Circuit analysis,

■ Software algorithm analysis by exercising individual routines,
■ Simulation at all levels of the design.

Embedded development teams and service designer will tailor these risk-countering activities to their special needs. For example, the embedded team might use automated flow diagram tools to indicate the actions occurring in the software and to define the sequences and branching in the code. Service designers can perform initial value stream maps to start squeezing money out of the process.

The team must perform design review using technically competent people in order to review design analysis results and design maturity and to assess the technical risk of proceeding to the next phase of the development process. The team should establish design review policies long before the final system design.

The most effective reviews are those that resemble a pool of hungry piranha as the participants attack the design under review with obvious relish. In short, reviews are poor reviews when nodding heads and snores dominate the period. Professionals should be able to investigate, prod, and question every aspect of the design. While it is not open season, the stature of the designer is irrelevant; the goal is to produce a better product. Groupthink during a review is dangerous and wasteful.

4.3.6.15 Design Release

One of the most critical concerns in the transition from development to production is the risk associated with the timing of design release. On some programs, design release schedules are established by back-planning from manufacturing schedules or ambitious marketing considerations. As a result, the design engineer is expected to meet unrealistic milestones forcing him or her to deviate from best design practices. The results are predictable—design solutions are not the most beneficial to the overall design, interface considerations are glossed over, expensive redesigns occur, and necessary documentation is sketchy. Expedited and advanced design releases create the need for second and third generation effort. On the other extreme, when the program team schedules a design release beyond the normal period required to complete the design, the designers may add complexity to the basic design rather than improve inherent reliability or maintainability or reduce costs.

4.3.6.16 Reducing the Risk

The team should

■ Identify practices and procedures for design drawing releases using documented corporate policies that simplify transition and reduce production risk.

- Flow down to other tiers the design release disciplines practiced by the highest supplier tier.
- Apply uniform practices and procedures dealing with technical requirements and evaluating current manufacturing capability and realistic design release dates.
- Alternate design approaches to help maintain the design release schedule in areas of high manufacturing risk.
- Validate complex designs before design release by fabricating preproduction manufacturing models (prototypes) and feeding results back to design for corrective action. This step increases the assurance that the design release documentation will support full-scale production.
- Include all necessary information in the design release documentation required for an orderly transition from design to production.
- Conduct a formal review of the design release documentation at a critical design review (CDR).
- Establish and validate a design baseline as part of the design release.
- Complete all design-related testing, including qualification testing, before design release to ensure that the design reaches acceptable maturity.

Integral to the development process is the fact that at some point, creative designs must then be released to manufacturing. The team and project manager complete the design release with the acceptance of the design through the CDR and qualification test process.

The entire process generalizes to both embedded development and service activities, particularly the risk mitigation behaviors. The automotive development tools are usable across the spectrum of products and processes (as we pointed out before, we know the documents in the food industry are similar to those in the automotive industry).

4.3.6.17 Product Requirements Risk

The product requirements are not the only things that add risk to the project. The team may poorly document the project requirements or even change them during the project. They can alter schedules and or change deliverables such as training requirements. Like uncontrolled modification of design requirements, these alterations can have a negative effect on the project.

4.3.7 New Equipment, Tooling, and Facility Requirements

Hardware investment (capital investment) can represent a considerable investment in the facility. Additionally, these may be long lead time items. An alternative strategy uses existing production equipment to accommodate the production requirements with group technology.

Clearly, these considerations are less of an issue with embedded development and service process design and implementation. The embedded developer may run into a situation where in-circuit emulation may not be available, but he or she still has the capability to program the microcontroller directly. Service processes often use people and not hardware to accomplish goals, so the need for substantial capital may not be an issue with services.

4.3.8   Gauges and Testing Equipment Requirements

Engineering and product specifications from the development phase are inputs to the production or process stage (the how-we-move-it-to-the-customer phase). The production and quality processes of individual manufacturers also provide boundaries or constraints for the measuring and monitoring system requirements (known, respectively, as gaging and test equipment in manufacturing). This documentation is as important to the manufacturing staff as product specifications are to the product development engineers. Developing these specifications during the product development process ensures the link between the design and process phases and involved people.

Some organizations try to fulfill these requirements with similar or existing equipment, cutting down on development and maintenance issues with the equipment. However, this solution is less than a win when the product and customer needs and expectations are unmet by this *recycling* approach. Saving some amount of time and money on the manufacturing line and processes at the risk of being able to deliver the product is a foolish trade.

4.3.9   Team Feasibility, Commitment, and Management Support

The team feasibility and commitment document provides management with the team viewpoint of the probability of success. This document details the costs of the effort, from project management to tooling and component cost. The team signs off and management reads the document and signs off also. This step secures formal management support for the project and provides for resources required by the team.

4.4   Cost

4.4.1   Request for Quote

For a request for quote (RFQ), the technical specifications must be complete enough that the suppliers getting the submission know on what they are bidding. In some cases, it may make more sense for the customer to specify performance rather than detail to encourage innovation on the part of the supplier. This approach applies just as much to embedded development and service design as it does to manufacturing enterprises.

4.4.2 Make or Buy Analysis

Besides the natural question, "Is this our core competency or how do we wish to invest our resources?", we have a much more tangible way to answer this question. If it costs more for the enterprise to make without differentiation in making it (we do it much differently or better than everybody else), then why make it?

$$Cost\ to\ build = Volume \times \frac{(Supplier\ cost)}{unit}$$

$$Cost\ to\ make = Fixed\ cost \times \frac{(Direct\ cost)}{unit} \times Volume$$

The interpretation of the results is found below:

■ If cost to build > cost to make, then *make*
■ If cost to build < cost to make, then *build*
■ If cost to build = cost to make, then *you choose*

This approach manages the financial aspects of the decision. However, if the area under consideration for make or buy is an area the organization believes is strategic to develop, then the strategic objectives may trump the immediate monetary gains. If the organization wishes to cultivate any real ability in this area (for example, project management), the activity cannot be outsourced to other organizations. The same can be said for tooling or any other function the organization believes is key to long-term survival.

Make the product, when:

■ We desire to have a high level of integration (vertically integrated),
■ Our knowledge of part/process is strategic to the long-term goals of the organization,
■ We desire control of production volume and quality,
■ We want design control (competition),
■ Our suppliers are not capable,
■ We have excess capacity

Buy the product, when:

■ Our suppliers have deeper pockets and access to technology that we do not have,
■ We have multiple suppliers for competition,

- We have no internal capacity,
- Our suppliers are more capable,
- Our knowledge of the part/process is not strategic to the long-term goals of the organization.

4.4.3 Value Analysis

Value analysis can be considered a subset of the acquisition process. The purchasing representative of the project team handles acquisition—an individual who may have the title *buyer*. It has been our experience that a typical acquisition is well handled by the purchasing staff insofar as it remains within the team information loop. A brief illustration is provided in Figure 4.7. The development of a cost-specific solution, on the other hand, takes more effort and thinking than simple acquisition.

4.4.3.1 What Is Value Analysis?

It is better to get the cost of the design planned than to try to cost reduce the design after the fact, unless downstream cost reduction is a strategic cost containment option. In some cases, a supplier is under substantial pressure to cost reduce a design/product by some percentage per year; hence, the supplier will sandbag (exaggerate) the value of the product by producing at the lowest cost at the beginning of the production run or by keeping the true cost unavailable to the customer.

Once a design has a tool to meet the design criteria, it is seldom effective to achieve a cost reduction by reworking the tool, unless a careful analysis reveals substantial savings from tool modification. Because the tool can be broken during its alteration, the supplier suffers an element of risk.

There are times when a launched design should receive a critique from a cost reduction team that is targeting cost reduction. This event occurs when there is a major change in technology or when system interfaces or components change (especially component obsolescence) or customer expectations shift.

MIL-STD-1771, *Value Engineering Program Requirements*, describes value analysis as "an organized effort directed at analyzing the function of systems, equipment, facilities, services and supplies for the purpose of achieving the essential functions at the lowest overall cost." The objectives of value analysis are the following:

- To provide assurance to the customers that they can purchase developed items for the most economical price during the product life cycle,
- To achieve cost reductions,

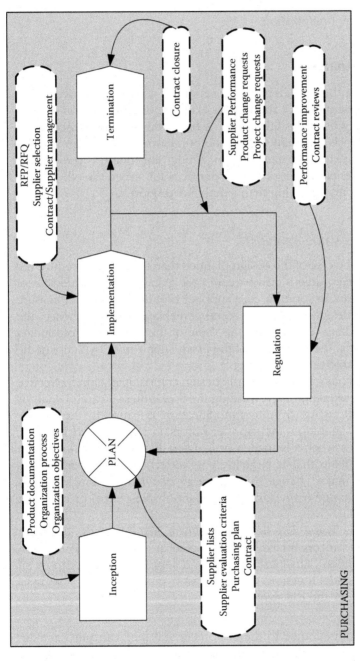

Figure 4.7 Acquisition oversight.

- To provide the buyer with an analysis of the cost (if the supplier and the customer have an *open book* relationship) and itemize what the supplier is doing to achieve the financial performance required for economical life cycle costs,
- To identify to the customer opportunities for value enhancement.

The single most important thing a project manager can do with respect to value analysis is to involve two major constituencies:

1. Upper management from their own firm
2. Project engineering on the customer side.

Organizations can spend a lot of time and resources improving the cost of a component or system after production starts, which increases the risk due to change. This situation occurs when the value team needs part modification in order to achieve price-down goals set by the customer. They are better off improving the efficiency of product production and expecting both supplier and customer to maintain a continuous improvement process—part of that effort should be devoted to cost and quality.

> The design stage is also the optimum point at which the vast majority of the cost of making an item can be reduced or controlled. If costs are not minimized during the design stage, excessive costs may be built permanently, resulting in expensive, possibly noncompetitive, products that fail to fully realize their profit potential.[3]

Many organizations perform some kind of value analysis. Value analysis is a critique of the implemented features and the proposed implementation solution. The review and analysis techniques improve the changes that maximum value will occur in the design. The cost considered is not only the material cost (although that is a significant cost source), but also the development cost (time and resources). This assessment can be a difficult proposition if the customer determines value.

There are two precepts to value analysis. One is based on the subjective concept of value. Value is dependent on the opinions, perspectives, and priorities of those performing the value assessment, often the customer. The other is based on the design functionality. The analysis considers the cost of the various aspects of the product, including but not limited to constituent parts, manufacturing, and assembly.

There are often five to six phases to value analysis. The discussion below focuses on the six-phase technique. Each phase has objectives that are built on and advanced by the next phase of the value analysis process.

- Information
- Functional analysis

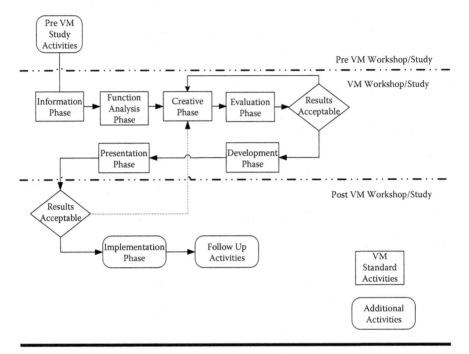

Figure 4.8   Value engineering process—SAVE.

- ■ Creativity
- ■ Judgment
- ■ Development
- ■ Recommendation

To provide the maximum amount of benefit, the value analysis process should be iterative. It should consist of a recurring review that critiques the existing product design solution to determine if the team can discover a more cost-effective possibility. The key to success is to understand the scope of the design; in short, understanding the function or functions of the product from the customer's viewpoint.

SAVE International defines the value engineering process as a six-phase process. The SAVE process for value engineering is illustrated in Figure 4.8.[4]

Target costing can be broken down into several functions (Cooper and Slagmulder):

- ■ Market-driven costing
- ■ Product-level costing
- ■ Component-level costing

Some factors that affect market-driven costing are the following:

- ■ Intensity of competition
- ■ Nature of the customer

- Degree of customer sophistication
- The rate of change of customer requirements
- The degree of understanding of future product requirements

Some factors that affect product-level target costing include the following:

- Product strategy
- Number of products in the line
- Frequency of redesign
- Degree of innovation
- Characteristics of the product
  - Product complexity
  - Magnitude of up-front investments
  - Duration of product development

Some factors that affect component-level target costing are the following:

- Supplier-based strategy
- Degree of horizontal integration
- Power over major suppliers
- Nature of supplier relations

What is the target costing triangle? The target costing triangle is a representation of the relationship of customers, suppliers, and product designers. The target costing triangle is similar in concept to the general product development triangle which is composed of quality, cost, and duration of development. The cynical reproach among software developers is that you can pick any two items from the product development triangle but not the third. We believe that the best way to achieve successful product development is to balance all three items; otherwise, costing will generally suffer the most quickly from uncontrolled quality issues and expedited timing.

The comments above suggest that scope is just as important when performing value engineering as it is in any other program management effort. Once the program manager loses control of scope to either a supplier or a customer, it is difficult to recover control and bring the project back on track. A value engineering effort is a project just as much as any other project and merits the same level of attention as any other project.

Frequently, the first effort of the development group—usually from the engineering department—is suboptimal for cost; in short, the launch team directs its efforts toward getting the product launched, into production, and into the hands of the customer. The expediency of this effort tends to eliminate consideration of optimal cost. That is not to say that cost was not a consideration from the beginning, but rather that time was brief and the engineers felt that getting a functional product into the hands of the customer was more important than optimizing the cost. Consequently, management often tasks the engineering department with attacking the product cost with a postproduction cost reduction effort. While it would be more cost effective for both customer and supplier to achieve optimal cost from the start of production, the cost reduction effort once the launch team achieves stable production is one of the more common approaches used by suppliers.

If the scope of the program is under rigorous control from inception, it is possible that the program manager will drive the engineers to optimal cost by the time they release the product for full production. However, the concepts of robust design—parameter design and tolerance design—can consume a significant amount of time and are often poorly implemented. We have seen projects where the parameter design and tolerance design occurred after the product went into production, not before. When the team skips these important steps during the development of the product, the product will generally have expensive warranty issues from the very beginning of production. Hence, we have two costs: the cost of some optimal design and the cost of warranty and subsequent customer dissatisfaction.

Given that the team may have to pursue a cost reduction effort after production, one approach would be to use the value engineering methods of Lawrence Miles. He broke the effort into three phases: identify a function, evaluate that function, and create alternatives to that function.[5] During the phase where the team identifies the function, information is the key factor, supported by avoiding generalities, getting answers from the best sources, and overcoming hurdles. While the chosen function is under evaluation, key factors are

- Having information supported by avoiding generalities,
- Having complete knowledge of the cost,
- Getting answers from the best sources,
- Overcoming hurdles,
- Putting a value on key tolerances,
- Using vendors wisely,
- Considering specialty processes,
- Using applicable standards.

In the third phase—creating alternatives—there are three key factors: creativity, judgment, and development. Several concepts support creativity: refining the function, using industry specialists, using vendors wisely, leveraging the skills of the vendors, using specialty processes, and again using applicable standards. What are the best sources? In some cases, these are the vendors themselves and their associated documentation. Frequently, the vendor has already performed significant test work on his or her own product. The goal, then, would be to leverage the cost-improved product from the work that is already completed by the vendor.

What about product selection for cost reduction? One approach is to sort products according to the amount of material content, either by dollars or by quantity of components. Another approach is to look at the high runners because these are the products that are most cost sensitive to even the smallest changes. Yet another approach would be to investigate products according to their gross margin, which picks up the value of material content through the cost of goods sold and reflects potential issues with direct labor and factory overhead. If a company issues reports displaying this information in tabular form, it can sort the tables to produce the desired ranking. We have found this to be the quickest method to selecting product candidates for cost reduction.

What about component selection for cost reduction? Component selection requires more profound engineering knowledge than simply sorting products by cost, materials, or margin. The choice of an incorrect component can lead to warranty issues, so it is critical that less expensive components receive appropriate testing either by the supplier or by the developer. Many suppliers support component engineering as part of their laboratory function. The duties of the component engineer relate to costing, quality, and the ability of the component supplier to deliver the component expeditiously. Additionally, the component engineer should consider the life cycle of the component and verify that the component chosen will become obsolete in the near future.

All of these tools will apply to embedded development and services also. They, too, can use a value stream analysis to assess the cost of a process, a tool available for the assessment of services and service cost. Embedded developers can find ways to squeeze more capability out of a given microcontroller, eliminating the need to move to a larger, more capable alternative.

4.4.3.2   Information Phase

The use of information acquaints the participants in the analysis with project goals, objectives, and constraints. The team distributes the product specifications and requirements for review. During this phase, the team will outline and emphasize the need (functionality), the cost (piece price), and cost (development) and delivery expectations.

4.4.3.3   Functional Analysis

Functional analysis is used to state the product features from the perspective of the end user (which can be the customer's customer). This declaration identifies the physical needs the product must fulfill without implying a technical solution, creating a functional target for the project team to meet, and opening the project up to the possibility of many possible solutions for investigation.

The same approach applies to process analysis. The team should develop as many alternatives to the existing approach as possible.

4.4.3.4 Creative Phase

This activity uses techniques such as brainstorming, TRIZ (systematic innovation), and other creative thinking methods to determine approaches for meeting the objectives defined in the documentation.

4.4.3.5 Judgment Phase

The use of judgment provides a critical analysis or review of the results of the creative sessions. The team uses the results to determine the ideas that are the most probable candidates for meeting the objectives of both the customer and the enterprise.

4.4.3.6 Development Phase

This activity turns the most promising ideas from the previous phase into usable solutions. The cost reduction team will add details to the remaining ideas for a more formal critique.

4.4.3.7 Recommendation Phase

This activity produces an output to the customer or project. In this phase, the cost reduction team considers the results of the development phase and presents the best way of meeting the objectives. Sometimes, the result is a palette of choices for the customer to select and prioritize.

What follows is a set of ideas that can be used for a value analysis study:

- Contract requirements
- Technical support
- Packaging
- Transportation
- Handling
- Data
- Schedule
- Hardware purchased
- Hardware built
- Customer-furnished equipment
- Manufacturing
- Policies
- Procedures
- Processes

- Tooling
- Test equipment/procedures
- Installation
- Equipment purchased
- Layout
- Operations
- Policy
- Staffing
- Maintenance
- Repair
- Cycles and levels
- Facilities
- Software testing

4.4.3.8 Product Functional Decomposition

Product functional decomposition is a technique focusing on the functional requirements of the product before concentrating on the form of the product. Decomposition means we take a top-level function and decompose (or break down) the function into its lower-level constituent functions. Another approach to functional decomposition is called functional analysis system technique (FAST). Value engineers often use FAST when trying to discover areas for cost reduction. Both embedded developers and service process designers can use FAST as an approach for cost-reducing their areas of expertise.

4.4.3.9 Value Analysis Integrated into Design Process

In the automotive industry, as in many other industries, predicting profitability for a given function is difficult. An experienced marketing department has the responsibility for assessing the probable value of the product and the price the customer is willing to pay. A supplier will often generate more effective costing if the approach to product value is systematic and quantitative (when possible) and with high-quality qualitative data when numbers are not available or only count data is available.

Cost rationalization after production start Many automotive and heavy vehicle corporations have a cost rationalization mentality; that is, the launch teams and product support groups must provide cost justification for development of the product or changes to occur. This approach is desirable given the competitive nature of business in general. Spending more time with the development up front to secure the best possible cost for the product is a much better solution than designing the product once, then cost reducing and the subsequent redesign, although the engineers can justify subsequent redesign when a technological improvement lowers costs significantly.

Cost effect   It is better to get the design to the lowest possible cost than to try to reduce the cost of the design after the development work and subsequent product production start because redesign risk is not present. We have all seen those curves that show the relationship between the stage of the product life cycle and any influence possible on the cost of the product. Unless there has been a radical change like those listed below, there should be little possibility for cost reduction. It is not surprising that calculating the product cost (while meeting all other product constraints for functionality and quality) sooner rather than later improves the profitability of the product for the supplier, the customer, and the consumer. In other words, the lowest cost at the outset eliminates opportunity cost issues that exist until the subsequent cost reduction. Some issues that can motivate a redesign are as follows:

1. Technological improvement in product component technologies
2. Technological improvement in manufacturing equipment
3. Process improvement in manufacturing
4. Major change in customer interacting components (systems and sub-systems)
5. Change in customer expectations

However, there is an alternative view. After the launch team completes the project and launches the product, the customer's purchasing organizations will request and require price improvements. They may have a contentious relationship with the supplier which leads to bullying to improve product cost. Often this cost reduction effort appears in the project contract and—in the automotive world—ranges from 3 to 6 percent annually. If the development team (or value team) designed the product to the least possible cost, these contractual targets will be difficult to reach, straining the customer and supplier relationship.

To ensure the best possible cost has been attained at production start and not have this postproject purchasing party around cost reduction, the following should happen: demonstrate the lowest price has been achieved in the design from the start so that the customer does not come back for a cost reduction every year, that is, educate the customer's buyers and managers.

Everything we have discussed in this section applies to embedded development and services. In the service process arena, activity-based costing may be a challenge, although Kaplan and Anderson [2007] provide a simplified solution. They use a departmental budget over actual time used approach to get good ballpark values for the cost of any activity of interest.

Quality   Companies can spend significant resources delivering a product to manufacturing with controlled quality. When the design function makes a change to the existing product or component, the product integrity function must expend some additional resources to verify the quality (appraisal cost). Many companies, having already expended the resources to deliver the product to market, will attempt to minimize the amount of resources used to certify the quality of a change to the product. This situation is true if the change is minor. Sometimes fewer parts are tested; other times a simple technical judgment is used instead of rigorous verification. For example, a stepper motor change to an instrument cluster appears to be routine and thus unworthy of further certification.

By eliminating appropriate qualification of change, the team increases the risk of releasing low-quality product, processes, or embedded software. In most cases, the value engineers should roll up the cost of qualification into the cost reduction effort to emphasize that the quality is not a secondary consideration.

Limitations   Cost rationalization after the design of the product places many limitations on the design staff and process, especially when these activities occur right after product release. For example, any observations that require tooling changes would be difficult to justify. Management may express concern in the form of asking why the product was not designed properly the first time.

4.4.4 Outsourcing

It is no secret that outsourcing is nowadays a major theme in business. The cost of delegating portions of the project to outsourced teams may be less prohibitive in other countries or other organizations may have competencies that are key to the project. Constant monitoring of these outsourced sites may be necessary (appraisal cost again). This requirement is true even when there are no language barriers or time zone difficulties.

Good business has always meant efficient use of resources and strategically developed competencies. The quality of informal communication between the project manager and the participants positively affects the project's success. Yet distributing these activities diminishes this communication. If the project manager distributes the project team, then he or she needs to consider what risks he or she incurs as a result and, perhaps, increase the travel budget to provide for the auditing that will be necessary.

When outsourcing, a clear statement of work and WBS is an absolute requirement. These work packages must be clear, intelligible, and attainable. The language in the document should be unambiguous and outline the deliverables from the supplier.

In some cases, enterprises will outsource embedded development to overseas development houses because these institutions are inexpensive and often have a high level of demonstrated capability. An example would be outsourcing to an Indian development firm with a capability maturity model (CMM) certification at level five, the highest attainable CMM designation.

4.4.5 Project Manager Involvement in Negotiations

Many organizations do not have the expertise to handle all of the work internally. Even when the capability exists, performing the task internally may not be cost-effective. The project manager should know the state of the relationship with the suppliers. Sometimes, a project may derail due to tough negotiations that produce delays in supplier deliveries. The project manager must know how these demands affect the relationship and any risk that may arise from these negotiations.

4.4.6 Modular Design

Modular design is not a new concept. It has been used within the software discipline since the advent of software functions. This technique allows for reuse of code modules, saving time and improving the quality of the software—library modules are present in disciplined libraries because these modules were used successfully in previous products. Extending this philosophy to hardware can improve the amount saved and enhance the probability of quality improvements. This method is easiest to achieve when the designing organization services similar industries; for example, automotive electronics. When the enterprise can standardize the hardware requirements (sometimes called "platforms"), they are available for immediate reuse. Sometimes the modular hardware designs produce difficulties—often the hardware has little commonality from project to project, especially with embedded design. However, it may be possible for certain core parts of the design to be modular, such as power supplies, and microcontroller or microprocessor sections and input/output to these devices.

4.5 War Story

4.5.1 Production Site

During the early negotiation phases of the project, the team had an idea to save cost by producing a component at one production facility in a specific geographic region even though the part would be used in both Europe and the United States. The European part of the customer organization wanted production solely out of Europe. The North American division, for logistical and performance reasons, wanted the product produced in North America. These protracted negotiations had an effect on the North American supplier's ability to secure funds for the creation of the production line. Ultimately, management determined that there should be two production sites, one in Europe and one in North America. This late decision posed serious risk to achieving the production start date. Management indecisiveness resulted in many long hours, numerous visits from the customer supplier quality assurance (SQA) staff, and needed a Titanic effort from the North American supplier to produce the product at the required quality level.

4.5.2 Material Availability

A product was being developed for a customer. Specifications from the customer determined functionality, cost, and quality targets. The supplier had difficulty finding the perfect microcontroller. The supplier preferred a controller under development as the new integrated circuit would have greater input/output capability and additional flash memory. The customer agreed and the project proceeded. In the course of the development, the microcontroller supplier decided it would not be able to make the new controller and the product had to be redesigned. The microcontroller supplier was unable to produce a product with stable flash memory and the collage of features initially advertised—it eventually sold a constrained version of this product.

The supplier's decision caused some developers to have to redevelop their product with new and more expensive controllers. The technology transitioned from the *leading* edge to the *bleeding* edge.

4.5.3 Battery

A particular tire pressure monitoring system required a battery that would be able to withstand the abuse of operating within a vehicle wheel for years. To determine suitability of the various battery manufacturers, the engineering and product integrity team developed a suite of tests to stress the various batteries available from different suppliers so as to ascertain suitability of the product. The team focused these tests on those areas identified to be the riskiest points in the design; specifically, thermal stress and battery life. These tests were independent of the product and were considered to be engineering level testing.

4.5.4 Commitment

A person was assigned a set of tasks within a project. This person was believed to have the requisite skills to accomplish the task and did not indicate that he could not perform as expected.

This person was assigned specification tasks (key tasks) within the project. He repeatedly said he could do the tasks and set the schedule (usually aggressive), and routinely missed the delivery deadline as well as the quality requirement. The first specifications came from him, were reviewed, found to be of inadequate quality, and were missing inordinate amounts of content of things he should have known. The manager worked with some other individuals to create a template so at least the headings would tell him what content needed to be included. He produced plagiarized documents complete with foreign language mixed in with English translations as the product specifications. These documents were inaccurate as well. When he was "called on it" during a meeting with the project development team and the manager, his reply was, "I do not want to be part of the team. I do not want to do this work." The corrective action was to remove the individual from the team, proceeding without the resource in the team (burdening the team that was already behind at least in part due to this nonperformance). The individual suffered no visible negative repercussion and when things were tight, he was the resource suggested to the rest of the team. This progression of decisions had a negative impact on team morale.

Chapter Notes

[1]Automotive Industry Action Group, Advanced Product Quality Planning and Control plan (APQP), (Southfield, MI, AIAG 1995) p. 13.

[2]Software Program Managers Network, *Project Analyzer*, (Arlington, VA, April 2000) p. 11.

[3]Donald W. Dobler and David N. Burt, *Purchasing and Supply Management* 6E, (New York, McGraw-Hill, 1996) p. 144.

[4]Value Methodology Standard (ed. 2007), (Dayton, OH, SAVE International 2007 ed.) p. 12.

[5]Lawrence D. Miles, *Techniques of Value Analysis and Engineering*, (New York, McGraw-Hill, 1961) p. 31.

Chapter 5

Process Development

A large part of this chapter relates to the embedded development process and service process design. Again, we feel that the automotive approach easily generalizes to most industries.

5.1 Delivery

5.1.1 Process Development Overview

The process development team (example in Figure 5.1) is comprised of all participants required to develop the production line. These customer and internal requirements include material handling, fault handling, work instructions, test equipment, and floor layout work. Customers may involve themselves in reviews and status meetings for the production line process. The illustration below is by no means definitive or all inclusive. Project demands and customer particulars dictate the staffing requirements. However, the graphic shows a minimum set of needs to fulfill the process requirements.

5.1.1.1 Phase Objectives

This phase is often run quasi-parallel with the product development phase. Sometimes the team will choose to start with a slight delay so it can adjust the process, see Figure 5.2, to the design. In service industries, the product and the process are the same. In this phase, the production staff works with the product design staff to produce the product at the desired volumes and quality levels defined within the specifications and contracts.

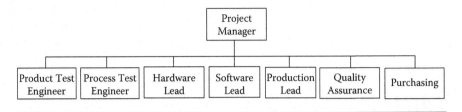

Figure 5.1   Example of a process development team.

5.1.1.2 Initiate

This inception phase can either appear within the overall project inception phase or can be a unique inception activity for this process phase. If we start a separate inception phase, the discussion includes only those topics related to process development. In this case, outputs from the product development and the previous phase, the voice of customer phase, are inputs to this phase. These inputs set the scope for this phase, which is to produce in volume the product of the project or to serve customers expeditiously.

5.1.1.3 Plan

The manufacturing part of the team launches the plan by considering those goals and risks associated with the process development or manufacturing

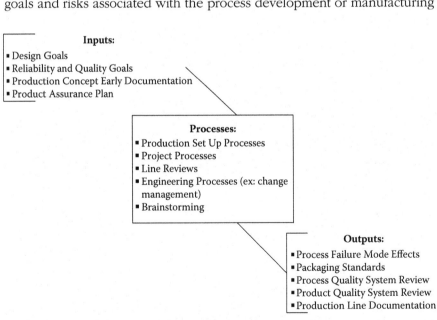

Figure 5.2   Process development interactions.

of the proposed product. This phase, as defined by AIAG, occurs in parallel with the product development phase; however, many organizations begin with some latency with respect to the design phase including following the design phase. Such action extends the development cycle. Delaying process development start reduces the risk of developing the manufacturing line around a product that is little more than a concept. Planning also leads to identifying those activities required to secure the quality of the deliverables.

5.1.1.4 Execute

During this phase, the team executes all activities and deliverables relevant to production. The project manager coordinates activities between those that are development-oriented and those that are process-oriented.

According to AIAG, the outputs for this phase are[1]

- Packaging standards
- Product/process quality system review
- Process flowchart
- Floor plan layout
- Characteristics matrix
- Process failure mode and effects analysis (PFMEA)
- Prelaunch control plan
- Process instructions

5.1.1.5 Control

This process consists of constant monitoring of the implementation phase and the subsequent deliverables. Control happens via action item lists, feedback from quality assurance activities, and comparative performance of selected metrics. When the team implements control correctly, output data feeds back to a monitoring function (can be the program manager) for real-time decision making. The process should be self-correcting.

The design and development of products and processes are subject to analysis for failure mode and effects analysis (FMEA) just as much as any other process (or product). The use of the PFMEA can be helpful for generating appropriate controls for the process.

5.1.1.6 Closing

The closing phase matches the expected deliverables and associated quality expectations to the actual deliverables.

Performing this work in tandem with the development work ensures that the design staff designs a product around processes that the production staff can fulfill. Gone are the days when the design staff *threw* the product over the metaphorical wall and the production staff was left to finish execution of the process. Parallel work helps certify the expedient introduction of the product. The risks from fast-tracking are that changes in an unstable design might affect the production scheme. When this situation happens, the work completed can recur, which often means more cost, more labor, and more time go into the process.

- Process requirements
- Process creation
- Production setup
- Reviews
- Production tools
- Product manufacturing handling
- Shipping

According to AIAG,[1] the outputs for this phase are

- Packaging standards
- Product/process quality system review
- Process flowchart
- Floor plan layout
- Characteristics matrix
- Process failure mode and effects analysis (PFMEA)
- Prelaunch control plan
- Process instructions

5.2 Product Integrity and Reliability

5.2.1 Process Flow

Any process can be documented with a process flow diagram. This diagram does not have to follow a standard (e.g., the ANSI standard for computer flow charts), but a standard may be used if it improves communication. The flow diagram provides a graphical representation of the process flow.

5.2.1.1 Ideal Process Flow

The first step in documenting a process should be a process diagram of the ideal. The ideal process occurs when a service or a product flows through the process with no errors. Start with perfection, create controls, and avoid planning for mediocrity.

5.2.1.2 Less than Ideal Process Flow

Once the team documents the ideal process, they can proceed to the second step, which uses the PFMEA we describe in the next section. The purpose of the PFMEA is to highlight areas in which the ideal process may fail. Once the team identifies a risk, it follows with a recommended action that results in adding a control to the process to prevent or detect aberrant behavior.

5.2.1.3 Process Flow under Control

The team has already created the process flow diagram and performed the PFMEA. Its next step is to add the controls to the ideal process flow (and any branches) so that it may move the real process toward the ideal process. Once the team knows the flow and the controls, it can document its process with a process control plan (the AIAG format may be used). Control plans are not unique to the automotive industry; for example, the food industry controls food quality and safety by performing a hazard analysis (analogous to a PFMEA) and elicitating critical control points followed by a plan.

5.2.2 PFMEA

PFMEAs are the process version of the FMEA used for design (DFMEA). The rules for executing a PFMEA are the same as they are for the DFMEA. Output primacy is still significant. Since it relates to a process, the output of any stage of the process contains the potential failure mode. The input or inputs of the process provide the cause. The effect, for example, is often something like sending an inadequate product down the line, but the team will typically express the specific item that has gone awry.

The team should formulate PFMEAs as soon as a line flow diagram exists. It should update the document every time the line design changes. Again, the goal is to head off problems before they become problems; otherwise, the production team will mire itself in corrective actions. Any process, including the process of project management, can be examined under the lens of the PFMEA.

When historical values are available, *Cpk* can be used to express occurrence in the form of PPM which, in turn, can be transformed into a value from 1 to 10. The AIAG added this feature in the 3rd edition of the FMEA blue book. If the team does not have a process under statistical control to use for *Cpk*, the more generic version, *Ppk*, is useful.

5.2.3 Key Control Characteristics

The customer may designate key or special characteristics as significant. However, the supplier can also derive or define these characteristics. Each customer has specific nomenclatures and symbols for designating special characteristics. Most of the time, the key characteristics derive from

- Variation unlikely to affect product
- Variation possibly affects product

- Variation probably affects product
- Safety compliance characteristics are always critical characteristics

The process control plan has a column for using customer-defined symbols to emphasize the key product/process characteristics. Note that this approach applies just as much to services as it does to manufacturing. The same goes for embedded development.

5.2.4 PCP

Process control plans are described in the AIAG Advanced Product Quality Planning (APQP) blue book. Basically, a process control plan (PCP) is a matrix of items listed in tabular/columnar format. Contained therein are

- Process number (to be tied to the PFMEA),
- Process/operation description,
- Device used for manufacturing,
- Characteristics,
- Special characteristics,
- Tolerances,
- Evaluation measurement techniques,
- Sample size and sample frequency,
- Control method,
- Reaction plan.

The purpose of this exhaustive table—the process control plan—is to reduce waste and improve quality during production. This approach accomplishes the goal through the rational introduction of controls; hence, the name "process *control* plan." The process development will identify special characteristics of interest to the customer with the customer-defined symbols. As with the FMEA tools, the PCP receives an update after any change to the process (making it a so-called "living document"). The team should define the PCP after creation of the flow diagram for the process. They may consider the PCP implemented when the quality engineers (or other appropriate designate) submit the production part approval process (PPAP) to the customer and the document receives approval. In the case where the process control plan for a service is used, the team can tailor the PPAP collection of documents to include only those items relevant to the service.

The header region for the PCP identifies the stage of production, number of the PCP, latest change level of the part number, part name and description, supplier or plant preparing the PCP, key contacts and phone numbers (be up-to-date!), the core team, the approval date, the origination date, the latest revision date of the PCP, the engineering approval date when necessary, customer quality approval date, and any other approval dates.

5.2.5   Special Process Characteristics

The team identifies these characteristics with such terms as "critical," "key," or "safety." This special identification emphasizes the significance of certain portions of the process. The significance may be due to the complexity of the subtask, customer demands, or potential harm to the employee or the product.

5.2.6   Process Floor Plan

The process floor plan is often a mechanically drafted to-scale drawing of the layout of the production line and associated activities (a map of floor movement may have value for services, particularly if the analyst of the process will do a value stream analysis). The team will take the number of work centers and work stations from the updated process flow diagram just as with the PFMEA and the PCP. The numeration scheme allows for a *relation* among these documents; in fact, some software packages support this relation by using a relational database.

5.2.7   Process Flowchart

The process flowchart documents the production process required to deliver the product or service; this chart is the formal version of what is called the "process flow diagram." This chart provides a description of each station and the processes required. Figure 5.3 provides a demonstration of the processes required to get a printed circuit board (PCB) prepped for the placing of components and subsequent soldering activity.

The process flowchart is one of the many steps it takes to produce any product. The sum of all of these processes illustrates the flow of the entire line. Additional information is contained in each of the following pages:

- Part information
  - Part number
  - Part name
- Sign-off
  - Design responsible
  - Production responsible
  - Quality assurance
- Component parts order list
  - Number
  - Part number
  - Part name
  - Supplier name

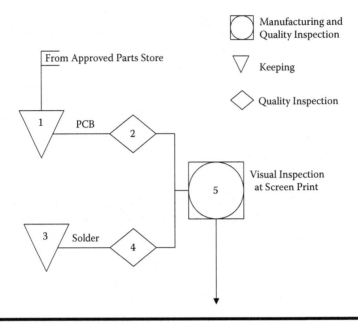

Figure 5.3 Process flow.

- Revision level
  - Date
  - Issue
  - Contents of revision (ECR)
  - Issuer
  - Checked by
- Symbol table

What follows is a short method for approaching this diagram:

1. Build a flow diagram of the ideal or near-ideal process.
2. Minimize the points in the diagram requiring a decision.
3. Analyze the flow diagram with a PFMEA (this step yields potential failures).
4. Provide controls for each significant failure mode.
5. Publish the final process diagram, with controls, to the appropriate team.

The short algorithm above provides a strong first-pass process.

5.2.8 Product and Process Quality System Review

In enterprises that fall under one of the International Organization for Standardization (ISO) quality standards, a product and process quality system review is mandatory. This review presents an opportunity to verify the control plan and update other supporting documents. The AIAG checklist for this review includes questions about personnel, training, various measurement system controls, visual aids, and many other activities. Service enterprises can use the same tactics to tie down their processes. The embedded developers can also benefit from this review.

5.2.9 Characteristics Matrix

The characteristics matrix links various product characteristics with the manufacturing operation responsible for contributing to the characteristic. This link allows for quick recognition of any change in these key characteristics after which it is possible to track the process that caused the change. Increased control of the process is the point of ongoing measurements.

In a service enterprise, metrics such as cycle times or wait times can have just as much significance as they do in manufacturing facilities. The "launch" team can measure the raw cycle time and derive model values from those measurements (we often model processes with arrival rates as Poisson processes and multiple Poisson processes at the kth arrival as gamma distributions).

5.2.10 Prelaunch Control Plan

The prelaunch control plan defines the system of control and other behaviors during the prototype phases of development. Early dimensional measurements, materials definitions, and functional tests (possibly by hand) appear in such a plan. Any control plan provides for containment of bad product/material and the prelaunch control plan is no exception. Given the benefits of this kind of control plan, it is surprising that neither of us has seen it implemented more than sporadically.

The types of controls seen during early production phases are much less formal than during full production. However, the general goal is the same—the protection of the customer. Examples of controls that are useful to this plan include the following:

- Frequent inspections (visual or preliminary semiautomated test equipment),
- Preliminary visual aids for operators or technicians,

■ Short-run statistical process control or, alternatively, *Pp* and *Ppk* assessments of the state of stability of the preproduction line,
■ Increased use of visual checks such as dock audits,
■ Substantial mid-management involvement.

It is not ridiculous to also use a preliminary PFMEA. This PFMEA can be used to place the controls in the correct places in the prototype process.

The idea behind a prelaunch control plan is that prototypes deserve the same level of attention as the final product. This control plan helps verify that the launch team is attentive to the product and the process in order to minimize the shipment of defective prototypes to the customer.

The preliminary control plan is also useful for pure embedded development and for service processes. In fact, the control plan approach is useful with *any* process. We use controls to provide an error-free process.

5.2.11   Process Instructions

We find process instructions placed at each station on the line, where they inform the operator of the specific tasks to use on the product at that work center. These instructions provide a sequential list of the actions to be taken by the operator at each individual station. It also provides information on potential problem areas for the station. The intent is to improve the repeatability and reproducibility of the production line. These instructions are not substitutes for training on the line in advance of production, but to provide support to the operator and to reduce unwanted product variation. Some examples of process instructions are

■ Short-order cooking instructions in a restaurant,
■ Assembly instructions for exercise equipment,
■ Firmware release information (usually comes with substantial installation guidelines),
■ Posters that provide the steps for setting up a personal computer

Note that none of the examples represent hard manufacturing. The use of process instructions is widespread. The use of controls, however, does not appear to be as pervasive.

5.2.12   Measurement System Analysis Plan

Measurement system analysis (MSA) is a tool typical of automotive corporations in particular and ISO companies in general. Supporting functions like laboratories come under the strictures of ISO/IEC 17025, which has more constraints than general ISO 9000 standard. The purpose of MSA is to verify the integrity of the enterprise measurement system. From a more pragmatic point-of-view, a true measurement system helps to reduce variation in the product and processes during manufacturing.

MSA has two major components:

1. Machine capability studies
2. Gauge reproducibility and repeatability (GR&R) analysis

We do machine capability studies to verify and minimize variation produced by the machine itself. Because part-to-part variation is of no interest, we can test the machine with a known good part to establish the quality of measurement (precision and accuracy). The team can also choose to test the machine with known bad parts to establish the limits of detection; they can also use a sample of good parts with some variation to measure the robustness to noise.

We perform GR&R analyses to verify repeatability of measurement by a given operator/technician; that is, the individual will measure the same part many times to establish the ability to perform the measurement itself. Reproducibility checks the measurement variation from one operator to another. We can use statistical software to determine the part-to-part variation, the within-operator variation, and the operator-to-operator variation.

As noted, a GR&R analysis provides information about part-to-part variation. We use analysis of variance (ANOVA) to ascertain the percentage contribution of the various components to the overall measurement variation. We can use regression methods also, but the standard ANOVA output is easier to understand.

Measurement is so important to the automotive industry that any customer can force a recall of a supplier's product if he or she can show that the product was manufactured with out-of-calibration instruments. Motor vehicles are complex machines where every deviation from target values degrades the overall result; hence, major efforts to reduce part variation are a requirement.

AIAG suggests that the supplier ensure the functionality/capability of duplicate gages, but the plan is designed to support the measurement system through development and production.

Services—for example, laboratory services—can perform GR&R as easily as any other operation. ANOVA is the weapon of choice again.

MSA, which includes both GR&R and machine capability, is impractical for embedded development unless the developers are using measuring devices to ascertain the behavior of the electronics. With embedded development, it is sufficient to use calibrated devices.

5.2.13   Packaging Specifications and Standards

We do not provide a complete discussion of packaging requirements. Most organizations have some person responsible for this aspect of the project. However, the project manager should have some knowledge of the needs and constraints to be able to ensure a suitable solution. In manufacturing enterprises, the team will often overlook the opportunities present in product packaging, for example:

- The packaging can provide advertising information,
- The packaging can provide special protection for the product,
- The packaging may be useful in cost-reduction analyses.

Some organizations specify constraints for material packaging. This specification accelerates processing at the manufacturing receiving areas. Organizations develop informed opinions over time on the best way to get product into their facility. Within these specifications, the team can execute tests for the proposed packaging to determine suitability to incoming material expectations. In other words, the team may ship test packages using various carriers to ascertain the suitability of the packaging.

Packaging specifications are determined from known information about the product design such as product sensitivity to mechanical shock, outside contaminants, and customer receiving and just-in-time (JIT) constraints and volume. In general, the steps in the packaging process are

- Concept
- Prototype
- Design review of prototype
- Testing
- Production

In short, the design of the packaging follows a similar process as the design of any other product.

Clearly, this packaging section is largely irrelevant to embedded development. Some services, such as UPS or Federal Express, use packaging as part of their competitive edge.

5.2.13.1   Corrugated Containers

Corrugated containers are useful for lightweight component shipping, but are less durable than returnable containers. The cost of corrugated containers is part of the cost of the product as a perishable item. The following items are considered when using corrugated containers:

- Handling labor,
- Handling equipment needed,
- Packaging composition,
- Floor space needed,
- Direct labor,
- Transportation cost.

In the twenty-first century, the *green* approach suggests that returnable containers are a form of recycling. Returnable containers are a good choice if they do not cause more consumption of resources through excessive freight usage.

5.2.13.2 Returnable Containers

Returnable containers are just that, returnable from the customer at the end of each shipment. This reduces and may eliminate the need for repurchase of shipping containers. However, when using returnable containers, the team should consider

- Costs
  - Initial
  - Repair
  - Transportation
  - Handling
  - Tracking
  - Administrative
- Volume
- Facility and equipment constraints
  - Storage
  - Floor space
- Return ratio
- Cleaning
- Volume
- Environmental hazards
- Product protection
- Inventory management

Suppliers should assess if any tax benefit appertains to the use of returnable containers.

5.2.13.3 Packing Slips

Packing slips provide information to the supplier and the customer of the container's contents. This item may seem trivial; however, when pressed to quickly place incoming material into the specified plant floor area or stockroom, it is easy to see why the accuracy of this information is critical.

Typical information contained on a packing slip appears in the list that follows:

- Supplier information
  - Supplier name
  - Supplier code number
  - Packaging slip number
  - Date
- Shipping information
  - Ship to
  - Bill to
  - Weight (gross, tare)
  - Bill of lading
- Packaging information
  - Number of units (quantity)
  - Unit of measure (weight, each)
  - Purchase order number
  - Program name/number
  - Customer part name
  - Customer part number

5.2.14 Preliminary Process Capability Study Plan

The preliminary process capability study plan ties to the prelaunch control plan and the final process control plan (PCP). The evaluation team can analyze preliminary processes with short-run statistical process control or *Pp/Ppk* assessment (see example in Figure 5.4). The *Pp/Ppk* assessment does not require that the process be in statistical control, but it does provide a pure statistical assessment of the status of the process. Note that the more common *Cp/Cpk* analysis uses the central limit theorem and small sample statistics to determine the level of control of the process (the tabulated coefficients used in *Cp/Cpk* analysis are dependent on the sample size).

5.2.15 Crashing Process Development

The processes referred to here are those that manufacture the product. Crashing this phase means, no matter how good the design is, there will be product risk when it is delivered to the customer due to insufficient consideration of production requirements. It is of little value to qualify the *design* if the team does not qualify the *processes* required to produce the

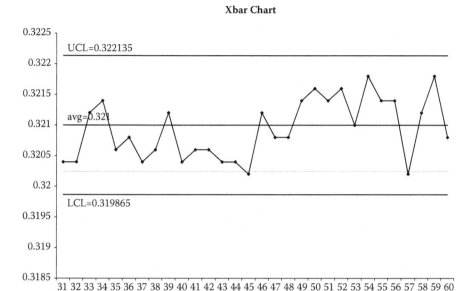

Figure 5.4 Control chart.

design. The result, poor quality to the customer, is the same. Simply adding more staff to an already compromised process will not necessarily improve the quality of the product. Beside the obvious diminishing returns aspects mentioned earlier, the actions that are taken to secure capability take time to execute. Some effects from crashing process development are

- Costs more to produce the line twice than it would take to produce the line once,
- Higher than expected warranty costs,
- High costs to retool the line because of missing needs,
- Downtime for the line effect on the customer,
- Duplication (another line to produce the product while corrections are made to the existing line),
- Insufficient production line verification.

We can generalize manufacturing consideration related to crashing the process development to service processes and the embedded software processes. The issue of product quality will assert itself at each opportunity. It is clear that any team crashing process development must progress with the utmost trepidation, using process controls whenever feasible.

5.2.16   Process Development Risk

5.2.16.1   Manufacturing Plan

Involvement of production and manufacturing engineering only after the design process is complete is suboptimal and represents a major transition risk. Consequences of late involvement are: (1) an extended development effort required for redesign and retest of the end item for compatibility with the processes and procedures necessary to produce the item, and (2) lower and inefficient rates of production due to excessive changes in the product configuration introduced on the factory floor. Increased acquisition costs and schedule delays are the result of this approach.

5.2.16.2   Reducing the Risk

As with any process, risk to manufacturing and the manufacturing plan can be reduced by anticipating failures. Additionally, we provide a list of support documents and activities that work in the automotive plant.

- Plan early while focusing on the specifics of the fabrication practices and processes required to build the end item
- Begin while the design is fluctuating and finish before the start of full-rate production.
- The following represent the key elements of a manufacturing plan:
    - Master delivery schedule that identifies by each major subassembly the time spans, promised delivery dates, and who is responsible
    - Durable tooling requirements to meet increased production rates as the program progresses
    - Special tools
    - Special test equipment
    - Assembly flowcharts
    - Test flowchart
    - Receiving inspection requirements and yield thresholds
    - Production yield thresholds
    - Producibility studies
    - Critical processes
    - Cost and schedule reports
    - Trend reports
    - Inspection requirements
    - Quality plan
    - Fabrication plan
    - Design release plan
    - Surge and mobilization planning
    - Critical and strategic materials
    - Labor relations

- ■ Manpower loading
- ■ Training
- ■ Training facility loading
- ■ Production facility loading and capacity
- ■ Machine loading
- ■ Capital investment planning
- ■ Make or buy criteria
- ■ Lower-tier supplier and vendor delivery schedules
- ■ Customer-furnished material demand dates
- ■ Work measurement planning
- ■ Energy management audits
- ■ Consider the following elements when generating a manufacturing plan—unique aspects of the acquisition may influence the plan.
  - ■ Colocate project staff and functional manufacturing people
  - ■ Build engineering and manufacturing test equipment with similar or identical systems
  - ■ Verify assembly planning before full-rate production
  - ■ Specify that a part of the design engineer's time be spent on the factory floor
  - ■ Combine assembly, inspection, test, and rework in unit work cells when appropriate
  - ■ Inspect development hardware with production line inspectors
  - ■ Build development hardware with participation from production people
  - ■ Develop the overall manufacturing strategy earlier in the purchasing cycle
  - ■ Verify the manufacturing plan and monitor progress against the plan with a series of contractual and internal production readiness reviews
  - ■ Include both prime supplier and lower-tier supplier in production readiness reviews
  - ■ Staff these reviews with knowledgeable people; that is, a mixture of manufacturing and design engineering people from outside the line organization doing the work

5.2.16.3 Qualify Manufacturing Process

The introduction of a recently developed item to the production line brings new processes and procedures to the factory floor. Changes in hardware or workflow through the manufacturing facility increase the possibility of work stoppages during full-rate production. Failure to qualify the manufacturing process before full-rate production with the same emphasis as design qualification—to confirm the adequacy of the production planning, tool design, manufacturing process, and procedures—can result in increased unit costs, schedule slippage, and degraded product performance.

5.2.16.4   Reducing the Risk

Qualifying the manufacturing process is in itself a process of reducing risk. The following list of ideas, tasks, and documents can help to mitigate qualification risks.

- The work breakdown structure, production statement of work (as identified in the contract), and transition and production plans should not contain any conflicting approaches.
- Plan a single shift, eight-hour day, five-day work week operation for all production schedules during startup.
- Adjust subsequent staff scheduling to manufacturing capability and capacity consistent with full-rate production agreements.
- Control the drawing release system:
  - Manufacturing has the necessary released drawings to start production.
  - No surge in engineering change proposal (ECP) traffic from design or producibility changes occurs.
  - "Block changes" to the production configuration are minimized. (A consistent configuration that does not need any block changes is an indication of low risk.)
- Minimize tooling changes and machine adjustments and ensure that the enterprise has alternate flow plans.
- Establish a mechanism that ensures the delivery of critical, long lead time items.
- Identify all new equipment or processes that fabricate the product.
  - Assign qualified/trained people to operate the new equipment and processes.
  - Achieve hands-on training with representative equipment and work instructions.
- Allocate hardware and other resources to proof-of-design models for data package validation and to proof-of-manufacturing models for implementation proof and production equipment troubleshooting.
- Qualify the manufacturing process at all tiers.

The manufacturing process required to produce an item influences the design approach and product configuration. Therefore, the manufacturing process is qualified with enough time for design or configuration changes to appear in the baseline product configuration before low-rate production begins.

5.2.16.5   Piece Part Control

Many automotive customers prefer to use standardized parts in their vehicles. This practice occasionally produces undesirable results when the standardization status of the part is more important than the quality of the part. For self-protection, customers should conduct intensive screening and inspection at their own facilities in order to provide an acceptable product to the production line.

5.2.16.6   Reducing the Risk

Piece part control can be enhanced with the following activities:

- Sometimes receiving inspection is more effective than source inspection; for example:
  - Suppliers may tend to ship better-quality products to customers performing receiving inspection rather than source inspection,
  - Receiving inspection costs are less than source inspection,
  - More lots per man-hour get inspection at receiving than at source inspection.
- Receiving inspection and rescreening exert contractual leverage on parts suppliers to improve overall quality of the product and, in the end, to reduce the cost of parts to the user.
- Piece part control includes provisions for screening of parts (especially mechanical and electrical components and electronic devices) to ensure proper identification and use of standard items already on the preferred parts list (a list of qualified manufacturers and validated parts).

A key element of parts control is an established policy that ensures that certain steps occur early in the buildup of the first hardware items to control part quality (both electrical and mechanical).

5.2.16.7   Supplier Control

Thanks to outsourcing and the fact that many companies have become wiser about their real expertise, reliance on lower-tier suppliers and on the skills of tier-one suppliers to manage their lower-tier suppliers has increased. The effective management of multiple tiers of suppliers requires a high level of attention, particularly as the logistics of shipping, inspection, and fabrication become more complicated.

5.2.16.8   Reducing the Risk

We can reduce some of the risk of dealing with multifiered suppliers by performing the following tasks:

- Answers to requests for quote (RFQ) can emphasize supplier management planning versus in-house management. Responses include the following:

  - Plans from a higher-tier supplier's organization for managing lower tiers.

- Plans for second-party audits of potential lower tiers.
- Tasks and associated plans to ensure that required up-front lower tier activities remain visible.
- Plans for program reviews, vendor audits, and production readiness reviews.

■ Prime suppliers conduct vendor conferences that address the following:

- Educate each lower-tier supplier on the requirements in his or her contract and the key elements of the prime contract.
- Communicate requirements to the lower-tier suppliers.
- Provide awareness of the lower-tier suppliers' role in the total system.
- Allocate resources to do the job right.
- Recognize and (when appropriate) reward good performance.

■ Higher-tier suppliers establish resident or frequent visit interfaces with critical lower-tier suppliers before production start.

■ Higher-tier suppliers maintain a list of lower-tier suppliers assisting personnel in emergencies.

■ Proper funding is committed to conduct the above guidelines during the early design phases to ensure adequate support to purchasing.

5.2.16.9   Defect Control

High defect rates in a manufacturing process drive up production costs because of higher rework and scrap costs. Product quality is a function of the variability of defects; that is, the higher the number of defect types, the lower the quality and vice versa. Lack of an effective defect information and tracking system not only increases production costs, but also degrades the product's performance in the field.

5.2.16.10   Reducing the Risk

The team should

- Identify types of assembly defects in terms of specific data, categories, and priorities for corrective actions,
- Track effectiveness of time-phased corrective actions,
- Monitor inspection and test yields and hardware throughputs always with predetermined action thresholds,
- Establish a feedback system to factory personnel and manufacturing supervisors,
- Reflect the criticality of defect information through factory policies,
- Monitor and track critical process yields to ensure consistency of performance.

A management commitment to defect prevention is the prime ingredient of a sound defect control program. A management policy on defect control is established during the development phase. This policy will require management involvement in the review of defect analyses and an emphasis on defect prevention that flows down to all lower-tier suppliers.

Defect control in embedded development is critical. The following ideas apply:

- Record all defects at every phase of development;
- Plot the defects as raw count and compare with a Rayleigh distribution;
- Check for defect containment; in other words, ensure that defects reported in a build do not make it to a subsequent build;
- Release the software when the Rayleigh distribution suggests the defect count is minimal.

5.2.16.11 Tool Planning

Tools are auxiliary devices and aids used to assist in manufacturing and test processes. They range from special handling devices to ensure personnel and equipment safety, to equipment required for methods planning to achieve the designed quality, rate, and cost. The risks associated with improper tool planning and proofing affect cost, quality, and ability to meet schedules. Poor tool control prevents workers from achieving desired production rates, failing to prevent or perhaps even contributing to errors in the build process and causing more labor to complete the task.

5.2.16.12 Reducing the Risk

The team should

- Document a tooling philosophy as a part of the early manufacturing planning process and concurrent with production design;
- Develop a detailed tooling plan to define the types of (hard or soft) quantities required for each manufacturing step and process;
- Include a similar plan for each subcontractor;
- Examine each tool rigorously before its introduction into the manufacturing process to verify performance and compatibility with its specification;
- Maintain strict tool configuration management;
- Establish and maintain an effective tooling inventory control system to ensure continuous accountability and location control;
- Establish and conduct a routine maintenance and calibration program to maintain tool serviceability;
- Colocate manufacturing engineering and tool designers with design engineers when practical.

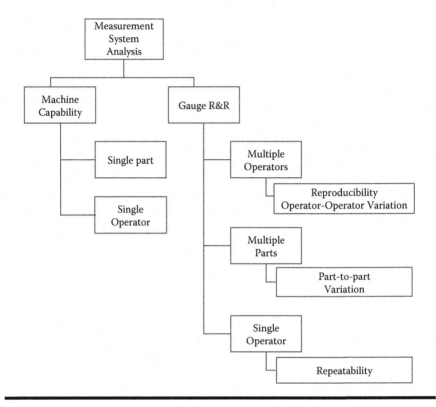

Figure 5.5   Gauge R&R.

Gauge R&R   Gauge R&R refers to repeatability and reproducibility of a given task on a production line (see Figure 5.5). The analysis is statistical and measures the effect of operators (or machines) on the process. Repeatability studies the ability of the operator or machine to perform the same way with the same material every time. Measuring with an operator is known as within-operator variation. Reproducibility measures the variation between operators or machines. When using an operator, we can call it between-operator variation. A customer or manufacturing facility may set standards for acceptable levels of variation on the line.

Gauge R&R should never be confused with machine capability studies, although they are related concepts. Machine capability measures the ability of the machine statistically, but is not a measurement of process per se. Machine capability studies are important because no process can perform better than the capability of its machines.

Poka-yoke   Poka-yoke means "mistake-proofing." The famed Japanese industrial engineer, Shigeo Shingo, felt that the mentality of control charts accepted bad parts when there was no need to do so. He reinvented the concept of mistake-proofing as a means of making it impossible to

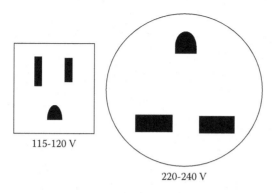

115-120 V

220-240 V

Figure 5.6   Poka-yoke.

manufacture a part incorrectly. Poka-yoke is usually inexpensive when used with production parts, with the most expensive components being the frequent use of limit switches that turn off conveyors when oversized parts push the switch.

Figure 5.6 shows a common household poka-yoke to illustrate the principle. The difference between a 120-volt connection and a 240-volt connection is sufficiently obvious to eliminate dangerous connections for all but those most interested in immolation of equipment or themselves.

Unfortunately, poka-yoke has no analog in embedded development or the software would be much easier to manage and release. The closest approach to poka-yoke is the use of previously proven software libraries.

5.2.16.13   Production Test Equipment

Special test equipment (STE, sometimes called "Peculiar Support Equipment" by the U.S. Department of Defense) is a key element of the manufacturing process. STE tests a product for performance after it has completed in-process tests and inspections, final assembly, and final visual inspection. Late STE design activities and the lack of the availability of qualified STE on the factory floor create unique technical risks. These risks include inconsistent final test measurements (when compared to test procedures used during the successful development program), false alarm rates that result in needless troubleshooting and rework of production hardware, and poor tolerance funneling that causes either rejection of good hardware or the acceptance of hardware with inadequate performance (type one and type two failures, respectively). Program consequences in this situation are schedule delays, increased unit costs, and poor field performance of delivered hardware.

Suppliers use production test equipment to filter out bad products before sending them downstream in the supply chain. Test equipment comes in many forms depending on the goals of the production line. We will describe a typical scenario in an automotive electronics enterprise.

With electronic parts, the test flow will look something like the following:

1. Top-side surface mount soldering (reflow)
2. In-circuit tester (ICT)
3. Bottom-side surface mount soldering (reflow)
4. ICT
5. Automatic optical inspection (AOI) (if available)
6. X-ray inspection (when using ball-grid arrays or landed groove arrays)
7. Wave solder (if necessary)
8. ICT
9. Manual assembly
10. Final assembly
11. Final functional tester

In this setup, we have three to five control points where the production test equipment is used to inspect the operating condition of the product. Indeed the automatic optical inspection is a robot that performs the same kind of visual inspection that can be achieved with human operators—with the added benefit of never becoming tired!

Some quality engineers have a problem with the idea of automated inspection and make comments about testing-in quality. The attitude is somewhat inane because electronics manufacture is not amenable to poka-yoke (mistake-proofing). Furthermore, when the supplier has many small boards to a panel, batching is common; in some cases, 80 boards per panel. The batching approach violates one of the most important concepts of *lean* manufacturing; namely, one-piece flow.

ICT at its worst is little more than an expensive method for generating scrap; at its best, it is a tool for improving line control on surface mount and wave solder equipment. ICT equipment will test—at a minimum—for open circuits, short circuits, missing parts, and wrong parts (note: these machines are sometimes called "MDA or manufacturing defect analysis"). More sophisticated approaches can infer problems when no electronic test points are available for test probes. A powerful ICT can detect design issues, borderline components, and programming of microcontrollers (special microprocessors).

In some facilities where parts control and furnace control is good, an MDA approach may be sufficient to ensure that no bad product flows downstream to other parts of the operation. At each stage, regardless of the approach, the goal is twofold: stop bad product and feed back information to the upstream operation for corrective action.

AOI is expensive and can have some of the same problems that a human operator would have; namely, the quality of the inspection is constrained to the quality of what can be seen. These inspection devices are used in low mix, high flow rate operations where the quality of solder joints is suspect and no human can inspect the material quickly enough.

Final functional testers will exercise the product through a preplanned sequence of activities designed to verify that the product is fit for shipment to a customer. With instrumentation, for example, a complete test can be performed on telltale lights, multiple gauges, audible annunciators, and a display screen in less than three minutes and often in less than forty-five seconds. Additionally, products with switches can be tested either electronically or with piston-actuation to physically push the buttons.

Commonly, test equipment is a *screening* activity rather than a *sampling* activity, meaning every product is inspected rather than a sampled subset of the product population. In our experience, customers tend to frown on approaches that include sampling, apparently with the naïve belief that screening will prevent a bad product from arriving on their dock. All of the test equipment mentioned above can be defeated by a weak solder joint (as long as the parts are in contact, it will have a closed circuit) with the possible exception of well-tuned automatic optical inspection, which is designed to detect that specific problem.

Also note that test equipment-style testing will only reveal problems with main effects while overlooking failures caused by component and subsystem interactions.

At the vehicle end of the production sequence, a variety of tests can occur. In some cases, the test equipment will verify dimensional characteristics using robotic feelers. In other cases, the vehicle builder will drive the vehicle over a test track designed to elicit failures from weak solder joints, poor connections, and other subsystems.

After multiple layers of test equipment activity, the end customer can still receive a defective part or vehicle. A motor vehicle is a complex system where failure modes *cannot* be mistake-proofed out of existence. The aperiodic (random) failure mode is a nightmare for every manufacturer in the automotive supply chain.

One interesting variant of the test equipment approach has been used by Toyota—it is called "autonomation." Autonomation is a hybrid form of poka-yoke, wherein some kind of sensor can detect an out-of-limits condition on a part and stop the conveyor or machine that is handling the part. The result of the stoppage is a high-profile emergency intervention by production engineers, with immediate correction and resumption of the process.

Embedded developers are fortunate if they can submit their software/ hardware for independent verification and validation (IV&V). An IV&V organization may have test equipment analogous to production test equipment, used for the express purpose of exercising the software to generate anomalous behavior.

5.2.16.14   Reducing the Risk

The team should:

- Develop a thorough factory test plan before detailed design of prime equipment;
- Require prime equipment designer input and concurrence on test requirements and test approaches;
- Test equipment engineers and maintainability engineers participate in prime equipment design and partitioning, test point selection, built-in test design, and design for test and maintenance and function;
- Colocate prime and test equipment systems design people when practical;
- Analyze the test approach for completeness of test and provide a feedback loop to correct test escapes;
- Employ a test tolerance strategy to catch problems at the lowest level, but do not cause excessive rejection of an adequate product; correct tolerance incompatibility with higher-level tests;
- Understand the capabilities of the prime equipment and use these to achieve simplifications in STE;
- Minimize attribute testing (binomial go/no-go type of testing) when it makes sense;
- Provide manual intervention capability in automated test equipment so that the operators can use the equipment while final software debugging is in process (this also can aid in debugging);
- Use breadboards of prime equipment, when appropriate, to begin debugging test equipment (this can enhance test equipment schedules);
- Assign equipment design people as part of the test equipment integration and verification effort;
- Allot adequate time for test equipment software debugging and compatibility verification;
- Require customer certification of factory test equipment and recertification if significant product and test equipment changes occur;
- Perform a thorough and realistic rate analysis to avoid shortages of test equipment (or overbuying); considered in this analysis are the number of expected failures in prime and test equipment in various phases of the program and equipment requirements to support qualification test, test, analyze, and fix (TAAF), engineering problem solving, and overhaul and repair;
- Use automated test techniques when rate requirements on the program warrant the investment.

The test team should design, qualify, and use STE as early as possible to ensure a uniform final product test from development through production transition. The STE design starts during the late phases of advanced development (that is, before any late milestones) and then the team should qualify the STE before full rate production.

5.2.16.15 Manufacturing Screening

Environmental stress screening (ESS) is a manufacturing process for stimulating parts and workmanship defects in electronic assemblies and units. When performed during development, it helps to ensure that the electronics hardware performs on demand, that the launch team knows the most effective screening levels before high rate production, and that the team discovers possible part type and vendor problems early. Do not confuse ESS with product validation testing (which is designed to demonstrate design maturity using production parts and the production process).

5.2.16.16 Reducing the Risk

The production team should

- Establish ESS procedures during development;
- Perform temperature cycling and random vibration when it makes sense;
- Perform random vibration because it stimulates more defects than fixed or swept sine vibration or similar levels of excitation;
- Perform dynamic testing—adjust procedures as indicated by screening results to maximize finding defects efficiently.

ESS techniques precipitate assembly and workmanship defects, such as poor soldering or weak wire bonds during the assembly process.

5.2.17 Process Capability

Measured   Process capability is a dimensionless measure of the six sigma ratio of the distribution against the specification limits of whatever we are measuring. Machine capability measures the variation of a specific machine against its defined specification limits.

Cp is the process capability and *Cpk* takes the process capability and relates it to centering between the specification limits. See other discussions comparing *Pp* and *Ppk* to *Cp* and *Cpk*. *Cpm* is an alternative that allows for a target value that may not be in the center between the specification limits (it is rare). The following formulae define the standard process indices in statistical terms.

$$Cpk = \min \left[\begin{array}{c} \dfrac{(\bar{X} - lower\,spec\,limit)}{3\hat{\sigma}\,\bar{R}/d_2} \\ \dfrac{(upper\,spec\,limit - \bar{X})}{3\hat{\sigma}\,\bar{R}/d_2} \end{array} \right]$$

Potential capability The term "potential capability" can be confusing. Some documents treat *Cp* as potential capability, since no centering is implied by the index. The AIAG uses a separate pair of indices called *Pp* and *Ppk*, which are the longer term analogs of *Cp* and *Cpk*. The idea here is that over some defined long term, a process will vary differently than over the short term. Indeed, there is so much variation that a standard Shewhart-style control chart will not pick up a 1.5 sigma shift well.

Limitation Most uses of *Cp*, *Cpk*, *Pp*, and *Ppk* assume a normal distribution; however, not all data is normally distributed. We will find nonnormal distributions in processes often if they are measured on an individual point-by-point basis. On the other hand, if the data is based on the means of samples, the Central Limit Theorem will drive the apparent distribution toward normality regardless of the underlying distribution.

5.3   Cost

5.3.1   *Cost and Delivery Performance*

Cost and delivery performance is critical to manufacturing, less so to embedded development and service processes. Clearly, the goal is to reduce cost versus price to increase the standard gross margin. In the automotive world, materials cost usually runs from about 50 percent of price to 65 percent of price. Other industries will have their own target margins.

We measure delivery performance in order to statistically calculate [Kruger 2005] the amount of stock we need to reserve as safety stock based on variations in supplier performance and customer demand. Failure to measure performance and take the appropriate actions to manage variation generally leads to stockouts and poor delivery performance (or its corollary, premium freight).

5.3.2   *Modular Design of Processes and Equipment*

The modular design of processes and equipment leads to quicker implementations, since the module is a byproduct of experiences with the equipment or processes. Also, these processes and equipment should already have the appropriate documentation to support full production.

Modularity in equipment and processes also suggests that the failure modes and effects documentation can be modular and that, in fact, is the case. A modular PFMEA or MFMEA saves documentation time.

5.3.3 Management Support

Management support is essential if, for no other reason, the executive management controls the finances for projects. This situation is applicable for manufacturing, embedded development, process design, and any other kind of project. The rule is simple: no management support, no project.

5.4 War Story

An automotive supplier dived into the six sigma approach to quality improvement, spending more than $300,000 in a two-year period to educate various types of engineers in the six sigma techniques. For the next four years, nothing occurred to produce any return on this investment.

Eventually, one of the American Society for Quality (ASQ) certified black belts put together a green belt class with some unique approaches: a week off to develop three cost-reduction projects, numerous exercises, training in detection of cost reductions and many more. In addition, the six sigma deployment officer developed a strong steering committee composed of the executive leadership for the supplier. Within a six-month period, the green belt candidates developed ideas for $1.5 million in cost savings for the supplier.

In short, a nonprocess generated no cost reductions. Once the six sigma program received full-time leadership, it blossomed into a healthy tool for promoting cost reduction and process improvement.

Chapter Notes

[1]Automotive Industry Action Group, Advanced Product Quality Planning and Control plan (APQP), (Southfield, MI, AIAG 1995) p19.

Chapter 6

Validation of Product and Process

6.1 Delivery

6.1.1 Validation of Product and Process Overview

Why are validation or verification activities performed? The list below identifies some reasons for testing:[1]

- Expose faults
- Demonstrate that requirements have been satisfied
- Assess the suitability of the product to customer needs
- Calibrate performance
- Measure reliability
- Make sure changes have not produced adverse side effects (regression testing)
- Establish a level of diligence that can be referenced in the event of product liability litigation

The validation development team (example provided in Figure 6.1) consists of a variety of people. Available talent should be able to devise the tests and create the program when and where it does not exist. Management verifies that the appropriate resources are available to perform the validation, although such is not always the case. Below is one example of a team that works with the project manager. In our example, we have numerous other players who are involved but not shown; however, a functional area individual leader represents these individuals. As with the other phases, the customer will often have a role within this phase, up to and

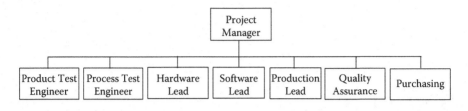

Figure 6.1 Example of a validation team.

including concurrent testing. Many customers perform audits or conduct a selected battery of tests duplicating the supplier's work and compare the outcome of the two tests for discrepancies.

Customer testing can be critical to embedded development, since the test group may not have the ability to develop sufficient verisimilitude to test the software product. Exhaustive validation of a product is impossible given the number of lines of code and the branching within the code. The alternative is statistically based testing and testing in the real environment.

The service team can test service processes by using real customers or simulated customers to challenge the process. Measurement will tell them whether they achieve their service goals.

6.1.1.1 Phase Objectives

During this phase, the evaluation team validates the product and the processes that produce the product (see Figure 6.2). In service businesses, the process is the product. This work starts much earlier in the process—the *specifications and development phases* (where the test plans are created) by the design and production test engineers. In this phase, the team will execute those test plans generated over the course of the project.

6.1.1.2 Initiate

The inception phase defines the scope of this portion of the project, just as it does in the other phases. The objective in this case is to ensure the product quality not only meets or exceeds requirements expected by the customer, but also meets or exceeds internal requirements.

6.1.1.3 Planning

The planning for this phase should start as early as possible during development. Many of the tasks are long lead time, meaning they require much time to coordinate in an effective way. The planning should minimize the unique risks for any phase of the development. Since the output of this phase typically happens so late in the project, managers will sometimes cut corners during validation—an action that raises the risk to product

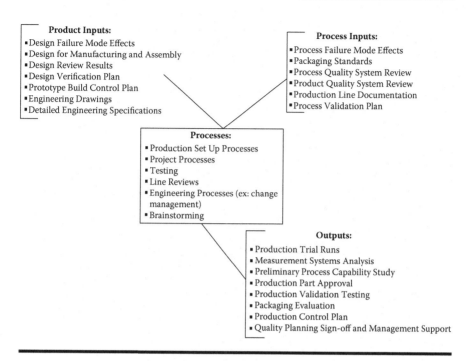

Figure 6.2   Validation phase interactions.

and process. This phase should start after the product or process specifications become stable. Long lead times for the creation of the test plan and the actual testing consume substantial time. For example, if the product is new and the team wants to automate some portion of the verification and validation work, the test team must validate the new approach. Every test validates both the test tool and the product or process, which can lead to undesirable oscillations between the testing and the tested.

6.1.1.4   Execute

The execution process of the verification phase consists of actualizing the assorted tests to prove that both design and process meet those requirements.

6.1.1.5   Control

Program control results from efficient monitoring of the output of verification and validation tests. Corrective actions are responses to failure reporting by the test team. A high level of control can be achieved by formalizing the failure reporting, analysis, and corrective action system (FRACAS) and making it a part of the quality management system.

Everything we say in this section applies to embedded development and process design testing with relatively simple tailoring.

6.1.1.6 Closing

According to AIAG,[2] the outputs for this phase are:

- Production trial runs
- Measurement systems analysis
- Preliminary process capability study
- Production part approval
- Production validation testing
- Packaging evaluation
- Production control plan
- Quality planning signoff and management support

Verification and validation requires a lot of time and money. If the project is in time- or cost-related trouble, the team may consider *short-changing* this part of the project—leading to increased risk.

6.2 Product Integrity and Reliability

6.2.1 Design Verification Plan and Report

Like many automotive documents, the design verification plan and report (DVP&R) in Figure 6.3 puts much information into a concise format. The format allows customer engineers and managers to review the status of testing using a brief format. It standardizes the reporting format and also provides a means for tying testing to the detection/prevention columns of the DFMEAs.

Some suppliers elect to pursue the American Association for Laboratory Accreditation (A2LA) honor of being accredited to the laboratory standard ISO/IEC 17025. This standard defines the quality system for both testing and calibration laboratories. ISO/TS 16949 (automotive version of ISO 9001) considers ISO 17025 accreditation to be sufficient evidence of compliance with 16949. The discipline required to achieve accreditation improves the quality of laboratory execution and brings the laboratory into a new community of excellence.

Figure 6.4 shows an example of how a component can be validated. This example shows multiple components receiving multiple tests. Each device under test (DUT) goes through a combination of tests. After each round of tests, the component is reviewed to ensure that the test has not adversely affected the appearance and functional performance before moving

System — 1 - Automobile
Subsystem — 2 - Body Closures
X Component — 3 - Front Door L.H.

DESIGN VERIFICATION PLAN AND REPORT

DVP&R Number	Revision Level	Key Contact	Phone	Date (Orig.)	Date (Rev.)
DVP&R 12345	DRAFT	Kate User	+1.520.886.0410	2/2/2004	2/16/2004
Part Number	**Latest Change**	**Supplier/Plant**	**Supplier Code**	**Engineer1 Approval**	**Date**
RS98765	Rev1	Acme Supplies/Tucson	RSS12345	Bill User	2/20/2004
Part Name/Description		**Drawing Number**	**Model Year/Program**	**Engineer2 Approval**	**Date**
3 - Front Door L.H.		DN333555	199x/Lion 4dr/Wagon	Jill User	2/23/2004
Core Team				**Engineer3 Approval**	**Date**
Kate User, Bill User, Jill User and Tom User				Tom User	2/23/2004

VALIDATION PLANS

Test #	Test Name	Test Method	Acceptance Criteria	Test Location	Sample Size	Start	End	Assigned To	Report #
1	Accelerated Corrosion Testing	Description of the accelerated corrosion test.	Corrosion less than XXX.	Test Lab A	Qty = X	2/6/2004	2/13/2004	Bill User	1A
									1B
2	Wax	Description of	Wax	Test Lab	N/A	2/8/2004	2/8/2004	Bill User	2A

VALIDATION REPORTS

Report #	Status	Start	End	Sample Size	Test Results	Completed By	Notes (Results)
1A	OLD	2/6/2004	2/9/2004	Qty = X	Description of the test results	Bill User	Notes related to the completed test results
1B	CURRENT	2/10/2004	2/13/2004	Qty = Y	Description of the test results	Bill User	Notes related to the completed test results
2A	CURRENT	2/8/2004	2/8/2004	N/A	Description of the	Bill User	Notes

Figure 6.3 Example of a DVP&R (Reliasoft xFMEA® output).

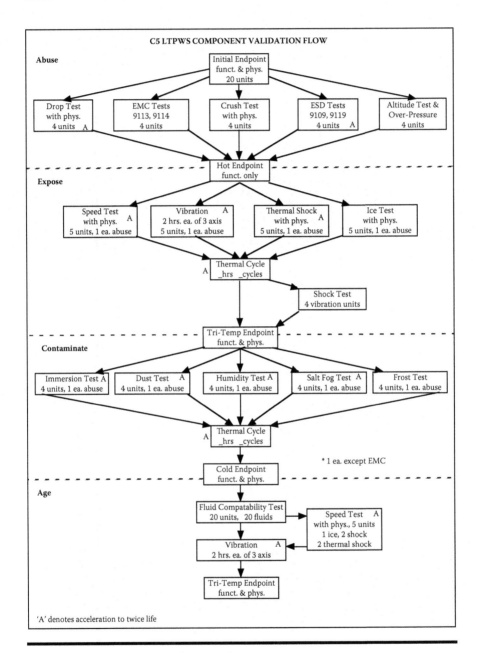

Figure 6.4 Example of a product test flow.

to the next test. This setup aids in determining the effect of the individual tests on the component, rather than trying to determine which stimulus in a battery of tests precipitated the fault. Endpoint reviews cover specific performance indices for the component in order to evaluate whether any damage occurred to the unit.

The verification and validation teams can better test through the use of designed experiments and multiple-environment testing. One-factor-at-a-time testing will only readily reveal egregious main factors and will only disclose interactions by fortuitous accident. The team should also see dramatic gains in time because it will run the different levels of the tests—the recipes—concurrently.

The designed experiment approach is available for embedded development, where it can be called "combinatorial testing." With embedded software, the evaluation team can stimulate digital and analog inputs on the microcontroller and record the results for comparison with the required responses. The same kind of efficiency occurs with software testing just as it does with hardware.

Service designers can also use the designed experiment to test the process against various stimuli. Attributed-based testing is available and powerful.

Project managers need to consider testing time as part of the development plan, particularly if testing reveals anomalies that force a redesign or a production workaround. The job of the laboratory is to find problems, not to prove that the product can pass an arbitrary standard. The project manager should expect the laboratory to report in the DVP&R format. Additionally, if desired, the project manager can take the dates from the DVP&R and transfer them directly into the project plan.

6.2.2   Verification versus Validation

The verification process ensures that work packages meet the requirements as specified, particularly given that the customer often specifies rudimentary testing in the specification. Verification weakly demonstrates that the work packages conform to requirements or specifications. Verification often includes activities such as testing, analysis, inspection, demonstrations, and simulations. This function is conducted on all levels of the design, from the documentation phases to bringing the physical aspects of the design to maturity.

An important part of verification is the peer review process. These peer reviews can be design reviews of the hardware and software. They can be physical inspections or structured walk-throughs. In software development, a powerful form of peer review is the *code inspection*, a formal, labor-intensive technique for revealing defects, sometimes performed earlier in the process than testing.

The purpose of validation is to demonstrate the product is capable of satisfying customer *needs*, rather than measuring the product or process against abstract requirements. A successful validation is confirmation that the product or process is ready for release.

Verification and validation activities often occur at the same time within the development process. Similar tools may be used to perform the confirmation. End users can perform some of the validation since they will often exercise the product or process in ways not considered by the formal verification and validation teams.

6.2.2.1 Verification

There are three aspects of testing:

- Software
- Hardware
  - Functionality
  - Appearance
- Durability

Most organizations have multiple levels of verification depending on the phase of the project. These phases are

- Engineering verification (EV) verifies product development theory
- Design verification (DV) verifies final product design
- Process verification (PV) verifies final production process design
- Continuing conformance (CC) verifies continuously the process and component

Each of these test phases may have specific and differing requirements. In the automotive industry, the DV and PV test plans are the same.

A simple example of verification activities by phase is given in Table 6.1. Typical failure classifications:

- Physical damage or quality perception
- Annoyance
- Functional failure
- Vehicle failure

Many organizations have their own qualification standards. In the automotive world, both supplier and customer can use industry standards from the society of Automotive Engineers (SAE), which attempt to quantify the physical environment for those electronic components. Still, industry standards and individual organizational standards are not the same, nor do they mix well. There are many organizations that believe their competitive advantage is due to the level of environmental stress their components

Table 6.1   Verification by Phase

Test	EV	DV	PV	CC
Extreme Temperature Soak	X	X	X	
Thermal Shock	X	X	X	X
Humidity Cycle		X	X	
Humidity Soak		X	X	
Mechanical Shock		X	X	
Drop		X	X	
Mechanical Vibration		X	X	X
Salt Spray		X	X	
Salt Water Intrusion		X	X	
Resistance to Fluid Splash		X	X	
Resistance to Pressurized Fluid		X	X	
Resistance to Fluid Immersion		X	X	
Dust		X	X	
Ultra-Violet Exposure		X	X	
Combined Environmental Cycle Test	X			
Combined Environmental Overstress Test	X			
EMC testing	X	X	X	
Connector System Testing		X	X	
Electro Magnetic Compatibility	X	X	X	
Noise and Vibration		X	X	
Switch Mechanical Life Test	X	X	X	X
Switch Contact Life Test		X	X	X
Vehicle Endurance Test		X	X	
Life Test		X	X	X
Test to Failure		X	X	

can survive. This may be true at times, but comes as a tradeoff with the cost of the design.

Verification testing can be time consuming and consists of functional tests, appearance reviews, and physical tests. As with many experimental environments, things can and usually do go wrong; for example, the complex test equipment may experience difficulties, and not be able to execute the required tests, thus having an effect on the project plan.

Many enterprises require testing early in the development process. Early testing helps ensure the design is capable of correct responses by the time of the design verification (DV) testing. This level of testing is what we call "engineering development testing." During this creation process, the design team subjects its design to various tests to verify that the design proposal meets DV expectations. This testing ensures the product development team is on the right track with design and component selection and suppliers meet the design challenges, more than just the analytical aspects of the development. It becomes confirmed that the computations and the supplier capability match the design requirements. The product receives several trials as the design progresses increasingly to the final design solution. As these test runs progress and more is learned about the design, the designers modify the product to reduce any design risk. At the last stage of testing, the component should have matured into the final solution presented for DV.

Example: new product development The design team is working on a battery-operated radio frequency device. The device has potting (filled with a moisture-resistant substance with the consistency of thick gelatin) to prevent moisture ingress. The device must be able to withstand prolonged exposure to high ambient temperature. The design team suspects the battery for the device may be a limiting factor in the design when encountering this thermal stimulus from both the high- and low-ambient temperatures required. The team searches for components that may meet these expectations, then designs a test regimen to verify the suitability of the selected components. This test consists of prolonged thermal cycling and thermal shock. Given our comments regarding designed experimentation, it should be obvious that this testing scenario is inadequate to reveal all but the most obvious design flaws.

Fault seeding Fault seeding attempts to qualify the number of faults within software, by deliberately and randomly placing faults within the software. The ratio of faults seeded to faults found represents the same ratio as total faults found and total faults within the software. The number of faults seeded and other specific faults are tracked. During the verification process, the number of seeded faults is monitored allowing for a ratio of seeded faults found to total seeded faults. This allows for an estimate of the number of faults that remain in the software.

1. N_s is faults seeded
2. F_s is the number of faults seeded that are found
3. F_f is the total number of faults found
4. N is the estimated total number of faults

$$N = \frac{F_s \times F_f}{N_s}$$

Pass/fail criteria Any test conducted must not only have the test requirements detailed, but specify success and failure criteria. While *pass* and *fail* are attributes, in general the product or process fails because it did not meet some measurable requirement. The test reports document the conditions for success and failure when issued following verification.

Importance project integration The typical resources required to test can be expensive. This situation is especially true for environmental verification (temperature, vibration, humidity, dust, etc.). The solution to verification and validation test equipment availability lies in planning of test iterations. MIL-STD-499(b) and IEEE-1220 recommend the use of a test and evaluation master plan (TEMP) as a way to accomplish integration of rational levels of testing with the project plan.

Test plan schedule The test plan for a component, process, or product often consists of environmental testing of the physical aspects of the design and functional and performance requirements. Ideally, all of the specifications for the product contain all of the requirements. This plan and associated schedule are important in order to synchronize sample availability with test equipment availability. The test plan should be devised in such a way as to provide confirmation of each of the required features from the specifications. Obviously, designed experimentation can take the supplier and customer beyond simple requirements verification/validation.

The test specifications are typically developed after the start of development activities. Developing the test plans in conjunction with the development of the product or process specification improves the chances of the product being verified and reduces risk. The requirement documents should have unique identifiers for each of the requirements. These unique identifiers for the requirements must have a matching test associated to verify the requirements.

Testing is important and an often neglected aspect of an embedded development project. There are occasions when testing concerns arise at the end of the project. If the testing is not considered during the development or it is conducted ineffectively, then quality risks for the project escalate. It is always much easier to solve problems or errors during early development phases than before the product or process makes it into full use. Once the product or process launches, the enterprise can expect increased cost for addressing performance anomalies, not to mention the negative effect on customer perception.

A well-defined test program will be a significant portion of the development budget. The testing becomes effective when adequate planning and organization occur and appropriate test techniques, design of experiments (DOE) are part of the strategy. Early test program development that coincides with product specification development work benefits the specification work by facilitating the identification and definition of the tests needed. Additionally, these discussions provide opportunities for identifying defects in the requirements and other design documentation. A well-developed plan helps the project team address observed problems when the cost of rework is low. Note that this situation applies whether we are looking at a product or a process.

The quality team should develop the test plan early enough in the process to find defects in requirements, design, and documentation promptly.

To some managers, testing sometimes looks like an expense to be cut when the project is over-budget. It is the duty of the program manager to defend the test budget. With nearly all but the simplest of products/processes, the execution of test plans provides the most effective means of eliciting defects and providing some confidence in the level of quality of the product and process.

Advanced enterprises can develop a TEMP that identifies and describes all testing activities and deliverables and significant milestones in development, particularly embedded software development. They should integrate this plan with the development work and the delivery schedule for the project. The resources and responsibilities should be understood and assigned as early in the project as possible. The identified resources must have the requisite skills and desire to perform these tasks. The TEMP helps to reduce this problem and should state the testing efforts and strategy for all project participants and management function to understand the need. The plan describes and assigns resources for all testing and evaluation activities. Use this plan with all members of the development team, including customers and managers. Program managers and those outside the project will find the TEMP helpful in relating the testing process to the overall project and its risks. The TEMP supplements the project plan and addresses management and technical issues associated with evaluation. As the project proceeds, the team will revise and update the TEMP as the project or program progresses to reflect current expectations, deliverables, and assignments.

In the case of software, good testing and evaluation occurs in tandem with software development. Testing is not just the phase that happens after coding. For maximum benefit and minimum risk, test development and preparation should complete before, rather than after, coding. Test engineering is more significant and integral to the analysis and design stages than it is to the coding and implementation stages. This is analogous to testing in school: the final exam may be important for demonstrating knowledge and for graduation, but it is the reinforcement provided by the testing and evaluation during the semester that helps students learn. In the case of a commercial program, testing is the main tool to learn about the product or process.

Pitfalls of testing The reason for testing is to certify that the component performs as expected under varying field-expected exposures. Capturing the field stimuli and creating a test plan that has good similitude can be difficult. In some cases, failures are not the result of a single stimulus, but of conditioning (preparatory events) before the testing or multiple stimuli to the component. Many automotive organizations, to gain better understanding of the field stresses on a component, will instrument a vehicle and measure the stimuli the component will often experience. They record these data for some duration for a range of vehicles and use the data to calculate what amount of time and level of these stimuli they should carry out on the component under study. They can use these preliminary data to derive test standards and validation requirements. Gaining a sample that represents the entire field with some statistical validity can be time-consuming and expensive if the expected failure modes have low probability, because low probability failure modes will drive the sample size larger in order to even see the failure.

Durability testing (life testing)   Many organizations specify durability or life testing requirements to demonstrate a certain degree of reliability at some statistical confidence. These tests must relate to failure modes expected to occur during the life of the product. Simply exercising the product without consideration to real field behavior and the physics of the failure will often produce a lot of meaningless data. Reliability testing has the following benefits:

■ Improved product robustness
■ The potential for improving reliability through a systematic reliability growth model
■ Early detection of significant design flaws through highly accelerated life testing
■ Prediction of product life based on physical models of the failure modes as observed during testing.

The project manager should verify that the engineers have related the reliability testing and the DFMEA. Again, the DFMEA becomes a key document for relating all parts of the product quality continuum.

Verification of prototype   Verification of prototypes can be a problem. In many cases, housings and other mechanical parts are models based on stereographic lithography (based on computer drawings and built from the deposition of a resin) to produce a model. These models have no mechanical strength and will not survive extremes of environment. Printed circuit boards are testable as soon as they are populated with parts. The purpose of *verification* is to determine whether the product meets customer-specified requirements. Frequently, the test organization is trying to determine whether the design will work.

What is important during this phase lies in the meaning of the testing. What are we trying to learn from this design? Sometimes the prototyping phase provides the most interesting tests due to the quasi-formed parts (same thing for a partially formed process). Many automotive organizations will use the society of Automotive Engineers (SAE) standards to define the testing. Beware! The SAE standards provide a firm baseline for automotive testing, but the standards are by no means exhaustive nor, in some cases, are they aggressive enough to tell the test organization much about the product.

European automotive manufacturers use the International Organization for Standardization (ISO) standards for automotive testing. These standards are, for the most part, analogous to the standards used in the United States.

When verifying a process, we often call the prototype build a *pilot run*, allowing the process development team to measure the new process while exercising it. The same questions of purpose and desired knowledge apply just as much to process testing.

Verification of final product   We call verification of the final product *validation* of the final product. Many project managers and engineers are unaware that the ISO definition of validation includes the concept of customer needs in addition to meeting requirements. This definition puts the onus on the supplier to test well enough that the supplier can say the product is robust. It also requires the supplier to have a thorough understanding of the customer product.

The AIAG calls this phase *product validation* and does not define it well. In general, the criteria to begin testing are the following: production tool, production components, and production process. End-of-line automated test equipment (ATE) does not affect form, fit, or function; hence, in some cases the products being validated do not need end-of-line testing before the validation begins. The purpose of this testing is to verify that the production product meets customer requirements and needs. If the supplier has an accredited laboratory, the laboratory management should be outsourcing to accredited laboratories.

Final process validation is analogous to final product validation. All of the expected subprocesses should be exercised as they would be under full production conditions. The approach is useful for nonmanufacturing processes such as sales and any other sequence of activities that can be defined to be a *process*.

For embedded development, everything we have discussed in this chapter is applicable. The primary issue with embedded software is the difficulty in testing the astronomical quantity of potential test cases.

6.2.2.2   Validation

In the automotive world, validation is the final set of developmental tests that occur subsequent to either the pilot run or the run-at-rate phase of preproduction. Validation occurs using production parts, a production process, and all operating production-oriented tasks. The goal of this kind of testing is to verify that the product the enterprise is manufacturing occurs correctly, it meets requirements, and it has not degraded the design through inept production.

Embedded software *validation* occurs before release of the software to production. If the embedded development follows the automotive companies' defect arrival rates with a Rayleigh distribution, they should *know* they are ready to release their software. Automotive companies will follow a disciplined, formal approach to releasing the software to the factory floor to avoid specious versions of the software making their way into production.

Validation and functional status class The functional status relates to the functional level of the product. Those functions that are not critical to safe operation do not require as much stringent and rigorous testing as those that ensure safe operation. Typical categories are

■ Convenience functions (least severe requirements),
■ Operational enhancements,
■ Essential functions,
■ Safety features (most severe requirements).

Dividing the functionality into these various categories allows the team to customize the testing per function.

Validation and test severity class It is not efficient to require all components to undergo the same amount of testing. Testing severity and duration is dependent on functional status; that is, the more important or critical the function, the more rigorous the required testing. For example, the amount and severity of testing for a home entertainment system are unlikely to equal the amount and severity of testing for brake and traction control systems on a vehicle. The severity class identifies the testing required to verify the component will perform in the field. The severity class concept is just as applicable to processes; for example, the failure to close during a sales process means no sale.

Typically, severity class also includes a definition of the expected performance from the component. We can define this performance by the following list (also applicable to process design):

■ Function is normal when test stimulus is encountered;
■ Function may deviate from normal performance when test stimulus is encountered, but will return to normal operation without intervention;
■ Function may deviate from normal performance when test stimulus is encountered, but will return to normal operation with intervention; and
■ Function deviates from normal performance during test stimulus and driver intervention has no effect requiring the component to be repaired or replaced.

6.2.3 Manufacturing Qualification

Manufacturing qualification often resembles design qualification testing. This suite of tests is used to confirm the capability of the production line (or process) to fabricate the design (or function) as intended. During this phase, the manufacturing line builds the product in the way proposed for production and often at the estimated volume. Then these components are subjected to a series of tests to verify that the production line is statistically capable. In this phase, the manufacturing engineers will check key product and key control characteristics to certify they meet the requirements.

Likewise, if we are selling a process (e.g., a turnkey process), then we would subject the components of the process to the same kinds of analyses. We will measure the capability of the process against desired results.

6.2.4 Electronic Data Interchange

Electronic Data Interchange (EDI) is the exchange of data between the customer and the supplier using a computer network and occurs computer to computer without human intervention. According to *Purchasing and Supply Management,* by Donald W. Dobler and David N. Burt [Dobler and Burt 1995], EDI is defined as:

> EDI is the direct electronic transmission, computer to computer, of standard business forms, such as purchase orders, shipping notices, invoices, and the like, between two organizations. In purchasing environment, documents are transmitted over the wire, eliminating the need to generate hard copies and to distribute them manually. By utilizing EDI, a buyer and a supplier are operating in a near real-time environment, which can reduce material delays by shortening acquisition times.

EDI implementation is not usually the domain of the project manager; however, it is the project manager's responsibility to make sure that the appropriate team addresses EDI concerns. Ideally, the project manager and the purchasing agent work together to make sure this aspect of the project works. This is how the material gets ordered for the customer's production. There are challenges around getting enough detailed and accurate information to the supplier regarding the volume which, in turn, generates orders on the supplier material ordering system. This can be critical at the start of the production since before production, the demand is quite low or does not exist at all. If this information were left alone, the supplier would order no parts and the launch would fail.

Note that EDI becomes more significant if mass customization is part of the supplier's competitive edge. In some cases (blue jeans, school buses, instrumentation), a supplier can fabricate enough *optional* features into a product to give the illusion of salable amounts of customization. Embedded software is important here also, because the embedded development team can design the software to be configurable.

6.2.5   *Bench Testing versus Field Testing*

6.2.5.1   *Bench Testing*

The first functional testing conducted on the proposed design solution is called "bench testing"—often given this name because the embedded software engineer conducts the testing at the design bench. It is rudimentary testing used to verify the performance of component functions in a way that ignores other possible system interactions. Verifying one feature at a time minimizes variables tested, but leaves out potential interactions. The design engineer is responsible for executing these tests. This person performs initial checks of the functions most recently developed and compares the results of the testing to the specification. If, in the view of the design engineer, the specification and the function are the same, this function is ready for the next round of testing and the design engineer will start coding other functions.

This level of testing requires a high level of expertise from the test team. The team should be working with complete test documentation, including an exhaustive list of expected stimuli and responses. Sometimes this kind of testing is referred to as *one factor at a time* (OFAT) testing. The major defect of this kind of testing lies in the inability to assess interactions with other portions of the code or other stimuli. The tester may be able to provide multiple stimuli to the product, but not to the degree seen in the end product.

6.2.5.2   *Final Product or Multievent Testing*

It is not realistic to believe that the software or even hardware operates decoupled from the system. Multievent testing can be quite extensive for a module of even minor complexity and feature content, requiring more sophisticated approaches to testing. Some examples from the automotive world should help make it clear what multievent testing is about. These approaches work for service processes and embedded development when the scenarios are similar. The common thread is the stress testing used to exercise the system whether the system is hardware, software, firmware, or a process.

Winter and summer test   Winter testing is a subset of standard vehicle testing. In the course of the development, the vehicle and components receive exposure to thermal extremes. These are in areas of extreme temperatures, such as cold parts of northern Canada and the heat of the southwest for companies in North America. In these climates, the vehicle receives driving stresses and may also undergo obstacle courses for a defined number of cycles (customer specific). This type of testing occurs before launch and will last during the development cycle with feedback from the test directed toward the project manager and the development team.

This portion of the testing can consume considerable financial and temporal resources.

Ride and drive   Often the customer will build up a prototype vehicle for engineering evaluation. These vehicles will be driven on a set of roads used for that customer's evaluation process. This drive will take the vehicle over various terrains lasting from a few days to a couple of weeks. This is a shakedown of the vehicle by engineers who evaluate performance of components with no intention of destructive testing. This testing is multi-vehicle with at least two engineers in each vehicle. They monitor various systems including data links and record driveability assessments.

If we expand the concept, we can use the same "test drive" approach to any service. We can execute the service against a known environment.

In addition to combinatorial and random (stochastic) testing, the use of a known environmental test suite has the benefit of being repeatable; that is, the evaluation team can compare its results with the results of previous tests.

6.2.6   Measurement Systems Evaluation

In the validation and verification phase, the measurement systems evaluation team reviews the results of the proposed measurement systems plan and subsequent actions. The real purpose of measurement systems evaluation is to ensure that measurements occur accurately; that is, the engineers are receiving a realistic assessment of what they are measuring.

6.2.7   Preliminary Process Capability Study

A preliminary process capability study will examine the process (whether the process is the product or makes the product) to see if it meets required values for statistical capability. In general, the engineers will assess the so-called *potential* capability and the degree of centering of the measured results against the mean value (sometimes the midpoint between the specification limits).

6.2.8   Production Part Approval Process

The production part approval process (PPAP) is a list of activities and methods required to deliver a component to production or service. The purpose is to ensure and verify the supplier is capable of producing the product to the design specification with no adverse effect on the customer's production line and end customer. The PPAP is designed to allow the production of parts that improve the supplier's production yield and thereby reduce costs associated with rework. This often concludes with the supplier quality assurance representative review of the functioning production line:

- Process flow documentation
- Work instructions
- Personnel
- End product audit
- Machine and process capability
- Gauge R&R review

A tailored variation of the PPAP is relevant for any process or product delivery. For example, the embedded software developers might issue a mini-PPAP before delivering the software and firmware to the test team, allowing the embedded developers a chance for review of their work and documentation before sending their product downstream.

We can also generalize PPAP to processes although the automotive list of required documents will often be irrelevant. The process design team should consider which documents reflect the completed state of the process before "releasing" the process as a final product.

6.2.9 Packaging Evaluation

Packaging evaluation for the product must happen early in the process and must consider the fragility of the product (see Figure 6.5). The team can develop tests to determine the vulnerability of the product. For example,

- Sensitivity or susceptibility of the product to shipping damage,
- Returnable material,
- Green requirements,
- Logistical chain,
- Default shipping method,
- Customer product use constraints (such as line sequencing or sub-assemblies),
- Customer storing constraints,
- Customer shipping cost tolerance,
- Volume of product and needs of production line.

A simple test of the packaging system could be to ship the product to the customer in the desired shipping containers. The customer would then inspect the incoming product with a brief evaluation of the externals of the containers. After this evaluation, the customer would ship the containers back to the supplier. The supplier would evaluate the containers and the contents for adherence to product specifications and shipping requirements.

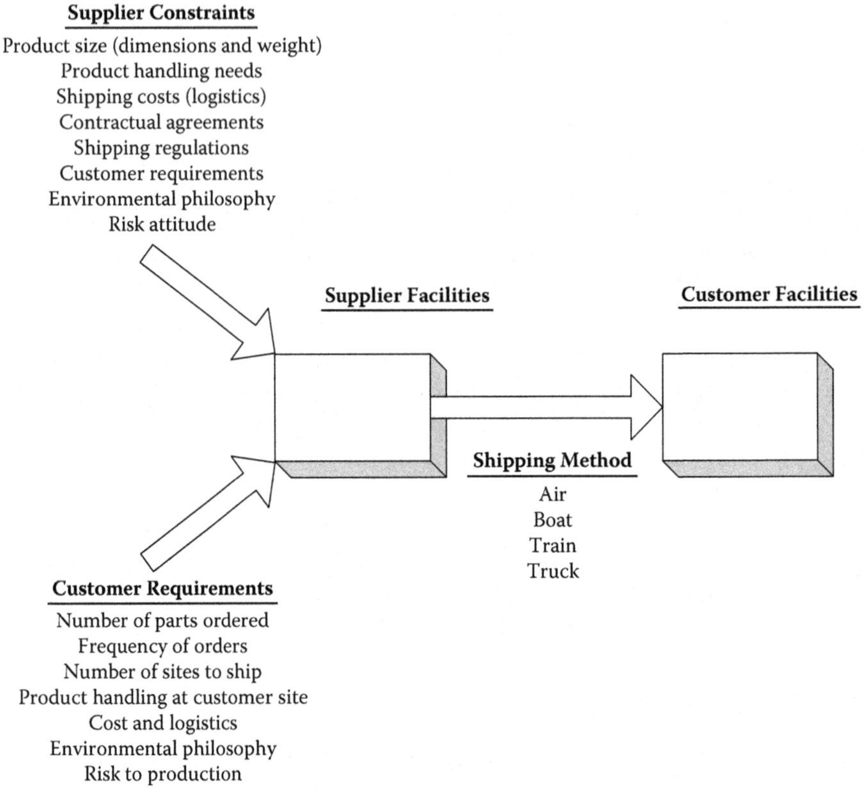

Supplier Constraints
Product size (dimensions and weight)
Product handling needs
Shipping costs (logistics)
Contractual agreements
Shipping regulations
Customer requirements
Environmental philosophy
Risk attitude

Supplier Facilities   **Customer Facilities**

Shipping Method
Air
Boat
Train
Truck

Customer Requirements
Number of parts ordered
Frequency of orders
Number of sites to ship
Product handling at customer site
Cost and logistics
Environmental philosophy
Risk to production

Figure 6.5   Example of a package and shipping drawing.

Additionally, standardized drop tests performed in a laboratory rate the damage resistance of the container.

6.2.10   Process Control Plan

We discuss the process control plan elsewhere. The reaction plan is the most relevant section of the control plan to this chapter. The reaction plan is the proven set of actions taken to keep the product within specification. If the team can anticipate points at which the process will venture into out-of-control conditions, we can also anticipate (and test) the steps necessary to return the process to control.

The idea of anticipation and problem solving is a core thread throughout the automotive approach to developing products. At every stage of a process, embedded development, process development, new product introduction, the teams can use the tools to eliminate issues before they happen.

6.2.11   Reduced Variation

Reduced variation occurs through the use of statistical process control (control charts) to determine the suitability and stability of a particular process. We represent variation with common cause (intrinsic random variation) and special cause (assignable cause variation). In manufacturing, the quality engineers will be responsible for maintaining product quality. However, given that the project manager is responsible for securing the quality of the resultant deliverable of the project, it is necessary that the project manager have more than a casual knowledge of quality processes.

A system with a predictable amount of variation is said to be within control. A process under statistical control is a process prepared for improvement and optimization.

6.2.11.1   Common Cause

When discussing variation, common cause is the variation introduced by ambient and unknown factors. These factors cause the output to deviate around a set point or mean value. In essence, common causes derive from the random variation intrinsic to the process. The team can improve common cause variation through a profound grasp of the process. Unintelligent process modifications are called "tampering"—tampering can drive a controlled process out of control. An alternative approach called "engineering process control" uses gains (adjustment values) to drive common causes toward the mean value and eliminate most of the intrinsic variation in the process.

6.2.11.2   Special Cause

Special cause or assignable cause variation is variation introduced by known causes and is induced by nonrandom input. Special causes are assignable (root cause and effect known) and are removable.

6.2.12   Crashing Verification and Validation

Crashing the verification and validation aspects of a project poses great risk to the quality of the output. This type of verification requires specialized equipment that is often expensive. To justify this type of investment, the equipment needs to have a high utilization rate, meaning ad hoc access to the equipment is unlikely. Further, crashing the verification and validation phase, generally at the end of the project, does not allow for sufficient time to make adjustments or administer corrective actions before the production start should anything be discovered during the verification and validation process. Which means, any production or design problem will make it into the field. Ultimately, these activities ensure the design meets customer expectations before the product is introduced to the customer.

6.2.13   Validation Risk

6.2.13.1   Built-In Test

Built-in test (BIT) circuitry in electronic products offers not only ease of maintenance in the field, but also more rapid troubleshooting during factory test and production. A source of concern—as with special test code—is the possibility of activation during product operation. Because BIT can improve cycle time during manufacture and fabrication, it is a powerful and cost-effective method for enhancing end-of-line validation of the product. Activation issues can be handled with exotic passwords.

A simple commodity example of BIT is the use of temperature- or voltage-sensitive battery testers in the packaging for household batteries. Smoke detectors have BIT capability to verify operation. Some emergency light systems function similarly.

6.2.13.2   Reducing the Risk

The team should

- Define maintenance and support requirements before inception of bit design;
- Provide design criteria for the contribution of bit circuitry to product risk, weight, volume, and power consumption;
- Conduct tradeoff analyses for each maintenance level on the interaction of bit, automatic test equipment, and manual test in support of fault detection and isolation;
- Conduct production design studies to define the use of bit in manufacturing inspection, test, and evaluation;
- Ensure that bit criteria, at a minimum, detect all process or product compromising failures and validate all redundant functions.

6.3   Cost

6.3.1   Outsourced

Outsourced testing is expensive. Automotive firms certified to ISO/TS 16949: 2002 must use laboratories accredited to ISO/IEC 17025:2005. When testing at outside labs, the test group must supply an engineer to journey to the external lab to ensure the test setup and execution matches requirements.

Any group designing a product should consider ramifications before outsourcing its product for independent testing. It is essential to keep control of the testing to certify that the product meets the needs of the customer and the easiest way to do that is to keep the product under one's own roof.

6.3.2   Simulation

Early simulations help confirm a course of action or a particular design solution. It is beneficial for the project manager to understand the benefit to project cost and risk mitigation. The mechanical engineers can use simulation for the mechanics of the design. Electrical engineers can analyze circuit components. For electronic components, some suppliers will supply models for their components. The component model creation simplifies the use of simulation tools and often represents a significant portion of the cost to simulation (especially when using PSpice for electrical components).

Process architects can simulate their process using a variety of commercial tools such as Arena® or GPSS®. Also available is the open source, agent-based tool called Netlogo® (http://ccl.northwestern.edu/netlogo/).

6.4   War Story

6.4.1   Last Minute Material Failure

A firm designed a product and determined that a liquid crystal display component would perform poorly in a high-heat and humid environment. The project staff brought this risk to the customer, who then decided that production would be delayed until a suitable corrective action to mitigate the risk occurred. Appropriate and challenging testing would have eliminated the component early in the development process. The closer to launch a major change occurs, the more likely such churning will lead to an inferior product.

6.4.1.1   Component Qualification

In a particular tire pressure monitoring system, the engineers needed a method of determining when the tire is above an angular speed to provide the appropriate monitor values. They found many suppliers who had components that appeared to perform the required end function. The desired tolerances and the overall durability of the switches were unknown. They developed a test method to assess the suitability of the various suppliers' offerings. This is an example of validation testing on a particular component of the design.

6.4.2 Trust but Verify

A new vehicle launch requires a new set of dash switches. The supplier has tooled a *standard* product for any original equipment manufacturer (OEM) to use. The supplier insisted that the switch had already undergone validation. A review of the validation plan suggested that limited redundant testing was in order. The test team devised a test plan consisting of a thermal stimulus and repeated cycling of the toggle switch. Eight hours later, they discovered significant failures. The switch was rated to endure thousands of cycles and the failure was within hundreds of cycles. Further investigation of the supplier's test approach revealed that the test was performed using a roller over the switch, which was nothing like the activation of the switch under actual field conditions. The verification testing, however, simulated the finger push as an operator would perform in the course of using the switch; hence, the detection of early-cycle failure.

6.4.2.1 PVT on Nonproduction Parts

A project was initiated for a company with multiple divisions. The resulting product was to be designed at one location and produced at the local production locations. The project was set up with this constraint, even though one customer did not agree to this constraint and cited reasons for not wanting to have production from Europe. Eventually, the project accepted the fact that there must be two production sites. However, this late change required securing funding at the local manufacturing site and put the production start schedule at risk. This local late production start put process verification testing (PVT) at risk of being completed in advance of the customer's production start date. The incomplete PVT was accepted to ensure material availability for production. In this instance, since the two manufacturing processes were common, the team accepted the risk of incomplete process verification testing. However, there were approximately four months of continuous improvement required on the production line to be able to produce to the desired quality and first-pass yield.

6.4.2.2 Slow Testing Feedback

It is not realistic to expect the development work to yield a fault-free product even when the quality controls are in place and with competent and motivated staff. Delayed testing feedback to the product or process designers can be deadly to project quality, budget, and schedule.

6.4.2.3 Errors Found in Production

It is often not possible to verify all combinations or permutations of a product or system. For very complex systems or systems that exist in multiple incarnations or variants, the combinations can be so high that there is no chance to perform all of the required verification. This can be especially true of heavy vehicle applications, particularly since there are often numerous variations as well as aftermarket systems sold for these vehicles. We have a number of experiences in which a variety of subsystems are put on the vehicle. Occasionally, these systems interact in an unpredictable way.

One particular combination produced an unpredictable response. Upon much investigation, it was possible to determine the combination of stimuli acting upon the control module to generate the failure. This sequence of events was described to the supplier. The supplier reviewed its software with this understanding and found an error in the code that would allow the problem to exist. The supplier was able to pinpoint the software malfunction rather quickly and made the necessary alterations to the software—and reverified. It is possible to conceive that if the supplier had a rigorous code review, with a critical eye, then the failure could have been found prior to production start thereby saving itself and its customer the required update of numerous vehicles.

Chapter Notes

[1] Robert H. Dunn, Richard S. Ullman, TQM for Quality Software 2E, (San Francisco, McGraw-Hill 1994) p174.

[2] Automotive Industry Action Group, Advanced Product Quality Planning and Control plan (APQP), (Southfield, MI, AIAG 1995) p25.

Chapter 7

Release to Production

7.1 Release to Production Overview

Release to production can exist as a phase or as a milestone, depending upon organizational processes. The project manager and the production team negotiate the end of the project. This date can be set during the project charter or as a result of the process development activities and identified risks.

The product hardware has been delivered and verified. From the embedded perspective, software release has occurred and is stable (verified and validated revision). We do not want to be going to production with software released immediately before start of production—we want to start our production in a stable, predictable environment. The processes for production have been completed and verified. The start of production (SOP) is eminent. It is time to release the product for production.

Release to production occurs when the supplier releases the product or process to manufacturing. Multiple releases can occur during this phase. These releases are tied to changes in the production, and are tracked with revision levels and associated dates. This happens as production hardware, software, tools, or processes are updated. Traceability of production line is as important as product software traceability. Production line traceability allows understanding production influences on the product in the field over time. If production-induced problems are encountered later, it is possible to understand the number of parts that are at risk. This is done by tracing lot numbers and product part numbers to revision level of the production line.

Usually, the project team receives notification that it has been *qualified* by the customer. This typically means the customer considers the product and processes ready for full production. However, approving the supplier

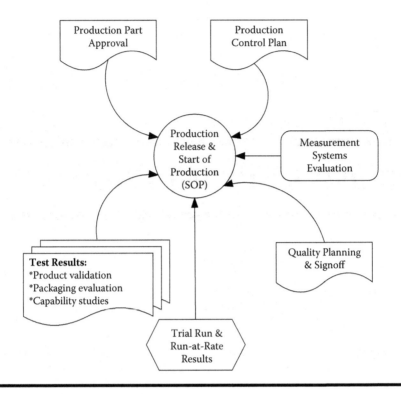

Figure 7.1   Release to production.

for production does not mean the customer has commenced production, nor that the supplier is fully prepared for production.

Often, the customer and supplier will verify that supplier-to-customer communications function properly. This is frequently done using Electronic Data Interchange (EDI) to transmit true demand and forecast orders. In some cases of release to production, the customer must contribute by providing real or realistic order data over a period of time in advance of the EDI. This is done to allow the material acquisition prior to production start or when the EDI is not available within the lead times for the hardware. Figure 7.1 shows one approach to production release.

7.1.1   Phase Objectives

The objective of this phase is to confirm the production line will meet the needs of the customer at production volumes. This typically consists of scrutinizing the production line under some stresses while observing the effects. The results are reviewed and any updates or corrections commence when required.

7.1.2   Inception

Contractual requirements and input from previous product and process development phases as well as organizational processes provide the scope for this activity. The product development and the process development work meet with the objective of producing a quality product in the volumes required by the contracting or purchasing organization.

7.1.3   Planning

The project manager and the project team determine how best to meet the production launch demands. When the production launch is not instantaneous, it will ramp up. Sometimes the ramp is slow and steady, at other times it progresses in fits and starts. In the early stages, the project manager must allow appropriate time for material acquisition.

Inputs to the planning process are:

1. Organization processes
2. Legal requirements
3. Industry practices
4. Contractual obligations
5. Customer production facility support needs
6. Product documentation
7. Initial production volume ramp up needs

This set of activities occurs near the end of the project. There could be considerable investment in the project to date. Any risk could have a heavy impact upon the project success. Identification of risks is required, perhaps more than in the earlier phases.

If the project's product is a service, it may be rolled out in stages or in particular markets. This launch may be put under some form of pressure (limit the resources or push capacity). This will generate feedback to improve subsequent introduction activities.

7.1.4   Implementation

The implementation process consists primarily of activities designed to bring the product into production. These activities typically take the form of

1. Process sign-off (PSO)
2. Trial production runs
3. Pilot runs
4. Methods builds (where applicable)
5. Runs-at-rate

6. Customer site product audit (product is part of a larger system)
7. Customer reviews of supplier manufacturing sites
8. Supplier reviews of customer product handling process
9. Customer volume reviews (EDI)
10. Troubleshooting and diagnostics manuals

7.1.5 Regulation

Control system regulation is possible via the results of these production activities compared to the desired or planned state. Discrepancies between the planned and actual results generate actions to bring the project to the desired conclusion. This situation requires identifying key metrics and measurement methods for determining any gap.

7.1.6 Termination

Termination is more than the satisfaction of contractual obligations. Termination happens when the contractual obligations have been met; however, this milestone does not necessarily mean the customer is satisfied. This situation can be critical as this is typically the last phase of the project. Project termination means future issues become a manufacturing and operational activity.

7.2 Delivery

7.2.1 Process Sign-Off

The process sign-off (PSO) occurs during or immediately after the run-at-rate for manufacturing processes. The team creates the documentation during the development of the manufacturing line and after recording the results of the run-at-rate. Typical examples of release to production documentation are as follows:

1. Part number and change level
2. Process flow diagram and manufacturing floor plan
3. Design FMEA (DFMEA) and process FMEA (PFMEA)
4. Control plan
5. Incoming and outgoing material qualification/certification plan
6. Evidence of product specification
7. Tooling equipment and gauges identified
8. Significant product and process characteristics identified
9. Process monitoring and operating instructions
10. Test sample size and frequencies

11. Parts handling plan
12. Parts packaging and shipping specifications
13. Product assurance plan
14. Engineering standards identified
15. Preventative maintenance plans
16. Gauge and test equipment evaluation
17. Problem-solving methods
18. Production validation complete
19. Buzz, squeak, rattle/noise, vibration, harshness (BSR/NVH)
20. Line speed demonstration and capability evaluation
21. Error and mistake proofing

If the team is releasing a *process*, they would tailor the document set to reflect the lack of a hardware/software product.

7.2.2 Trial Production Runs

The manufacturing team reviews the development of the manufacturing line with a trial production run (TPR, sometimes called a "pilot run"). The TPRs happen before the run-at-rate or production part approval process (PPAP) reviews. The team uses these trial runs to identify problems in the production line undiscovered during design. In these reviews, the team recommends areas for improvement and modifies the line based on empirical data, comparable to the way professional football teams use exhibition games to adjust their strategies. The approach generalizes to any process.

During the TPR, there are reviews of the tools created for the line and those that are still in progress. The quality assurance engineer leads these reviews. However, it is often beneficial to have some portion of the developmental engineering staff present from both the supplier and the original equipment manufacturer (OEM). These critical, multiperspective reviews help ensure the end product from the line will meet quality requirements and customer expectation. This review is especially beneficial for lines that employ large amounts of manual labor due to the human-oriented quality of the controls (visual inspection). The reviews produce improved work instructions and tools for the assembly process since poor instructions and tools become apparent during the run.

The equations show key metrics for either a service or a production process, namely "first-pass yield," which measures the output of the process before the application of any correction. First-pass yield reveals the quality of the documented process. Rolled throughput yield is the sum of "Y" individual throughputs (from each work center), which generally leads to a lower value than first-pass yield. Both equations represent measures of the *correctness* of a process.

$$First\text{-}pass\ yield = \frac{(\#\ built\ correctly\ without\ rework)}{(\#\ pieces\ planned)}$$

$$Rolled\ throughput\ yield = \prod_{i=1}^{n} Y_{TP_i}$$

7.2.2.1 Production Process Review

The production process review employs a critique of the production line under stress. The launch team can induce stress by speeding up the process, by removing labor from the process, or by introducing deliberate errors and observing the reactions. This is often done with the customer witnessing the build.

7.2.2.2 Process Verification Testing

Parts from this run become parts for process verification testing (PVT). Often the PVT is the same test regimen as the design verification. It is critical to consider the number of parts required for both this production run and the subsequent verification activity. A small sample makes assessment of the production process capability speculative at best. Additionally, the yield or number of useable parts from the production build have an impact on the number of parts available for the testing activity. For example, the PVT activity requires 50 parts. Unless you expect your first-pass yield to be 100 percent, you will have to build more than 50 parts (see first-pass yield). Many customers will not allow reworked parts to be used in the PVT activity.

The product integrity function performs tests on the end of line product to ascertain the quality of the build. Since the team has already verified the design earlier in the development process, it is now verifying the product as produced under realistic conditions. This requires the line to be representative of the final production.

7.2.2.3 First Time Capability

First time capability measures the capability of the lines to produce the product at estimated quantities. First time capability is a misnomer in the sense that the production line has not yet proven to be stable; hence, any calculation of stability has the risk of being nonrepresentative.

7.2.2.4 Quality Planning Sign-Off

The launch team compares this production run to the documentation for the production line. The team compares the documentation (such as control plan, process flowcharts, and process instructions) that articulates how the line should work by design to the actual performance of the line. Sometimes, labor does not follow the designed process, which requires corrective action; sometimes labor is not capable of meeting the demands set upon it by the designed process. This is addressed with tools or automation. Other times the team modifies the documentation to reflect the actual methods as witnessed.

7.2.3 Pilot Runs

7.2.3.1 Goals

The goal of a pilot run is to verify that the production equipment, gauges, fixtures, and processes are running. The pilot run does not provide statistical data for calculation of process capability nor can the results be used for PPAP submission to an automotive customer since realistic run rates are not achievable during a pilot run.

7.2.3.2 Objectives

When the launch team executes a pilot run, the manufacturing organization should audit the process with a production readiness checklist. The checklist helps set a minimal level of expected behavior and is also an efficient way to record the results. Military standard MIL-STD-1521B presents a high-level list for the production readiness review.

7.2.4 Runs-at-Rate

There is significant risk in the first production run volume from the early production line setup. During these activities, the team stresses the production line to produce the amount of material per unit time as it would expect for the line to meet the customer's expected volume of product. This stress in the production line and associated processes reveals risks and weaknesses. Solutions can then be found earlier in the process. A competent run-at-rate analysis provides the following:

- Corrective actions to happen before actual production in order to avoid having an effect on the customer's production line
- Comparison of actual to theoretical line throughput
- Constraint identification and management

■ Feedback to improve line efficiency
■ Capability demonstration (measuring centrality and dispersion of data)
■ Proficiency improvement of the line operators and technicians before full production

7.2.5   Crashing Production Release

We have already discussed crashing the project and related difficulties. A release to production when not prepared poses risks not only to the launch and to customer satisfaction, but also to the resources and money to correct subsequent field failures; for example, added maintenance intervals or product campaigns and recalls. These latent failures are often not discovered for months to years after production volume has been established.

7.2.6   Methods Builds

As the product comes to the production level, parts are shipped to the customer's manufacturing facility. Figure 7.2 shows how the methods builds are used for *fitment, material handling,* and *installation trials.* The launch team uses these parts to test its ability to build the vehicle (or subassemblies), allowing for a determination of how best to get the part into production and an estimation of the time to put the parts on the vehicle. This effort helps determine what the launch team requires in order to put the product into the vehicle and material staging demands.

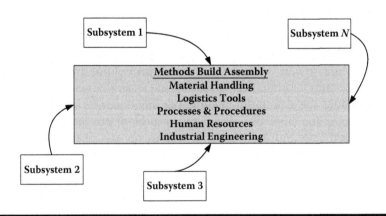

Figure 7.2   Methods builds.

7.2.7 Design Change Notification

Design change notification (DCN) is one way engineering communicates changes to the manufacturing system. DCNs apply to both supplier and customer. Driving engineering changes into the manufacturing process means controlled introduction of new or updated designs. Changes as trivial as changing values of capacitors or resistors require updates and alterations to the pick and place machine (surface mount soldering), conformal coating (circuit board protection), or other devices. Furthermore, the customer must evaluate each of these changes for effect on the form, fit, function, or quality (risk). A proposed change that has an effect on any of these areas will require a DCN to initiate, log, and record the change. Additionally, a change of this magnitude usually changes the part number of the product. An example of the need for a DCN is found in Figure 7.3. The farther along the team is in the process, the more costly the changes. Design changes can occur at any time during both product and process design activities, wreaking havoc on budgets and schedules in the later stages. Change control is significant to all phase of any project. However, release to production (or process) means the team is in the end game. Time is running out on the ability to make changes before the customer expects the final product. The shorter the time to implement the change, the more pressure to bypass the process.

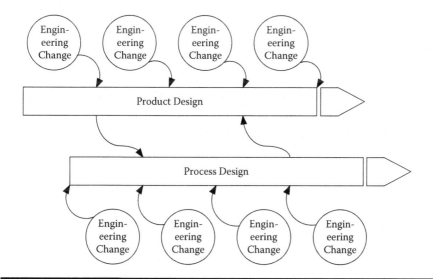

Figure 7.3   Design change notification.

7.3   Product Integrity and Reliability

7.3.1   Production Test Equipment

During the development process, the launch team identifies production test equipment needed to ensure product quality. This equipment is often a significant part of the cost of tooling the production line. This expensive equipment may include vision testers or *bed of nails* circuit board testers (if the product is electronic).

If we are dealing with a service, we may be able to test the service by using simulated customers during the process design phase. Even after we launch the service, we can add controls that provide a measure of the service, even if those are solely in the form of questionnaires.

7.3.2   Production Release Risk

When the team releases a product to production, any potential risks evolve into real risks. If the design for the production release went through a comprehensive process with no omitted steps, one probable set of risks lies among the support release activities. These include but are not limited to

1. Service parts,
2. Technical documents (product support manuals),
3. Product documentation (deviations),
4. Aftermarket and service training.

Sometimes these activities are neglected or carry a lower priority. When one is behind schedule or short-handed, these activities do not help secure the actual product release. Producing the technical documentation and handling the service parts becomes a secondary activity. However, these activities are just as important. Getting this wrong or poorly delivering the technical documentation for the product has a ripple effect on all downstream activities. If it is not possible to troubleshoot parts, the product will get a poor reputation which will stick to the customer and affect the supplier.

7.3.2.1   Service Parts

Service parts become an issue when unplanned. They have different packaging requirements than the standard product. Because they are service parts, forecasting becomes difficult due to unpredictable part failures. Service parts frequently go to a dealer and thus cannot use the packaging and shipping plans of the production version.

7.3.2.2 Reducing the Risk

The following provides some ideas for reducing risk for service parts:

■ Develop a service parts purchasing strategy early in final system design to identify least-cost options.

■ Use the same quality manufacturing standards and risk reduction techniques for service parts manufacturing and the repair process.

■ Plan the transition from supplier to customer service parts support on a phased subsystem-by-subsystem basis.

■ Base initial service parts demand factors on conservative engineering reliability estimates of failure rates.

■ Consider plans for developing service parts acquisition and manufacturing options to sustain the system until phaseout, particularly with respect to phaseout of unique raw material items. These plans include responsibilities and funding for:

  ■ Configuration management,
  ■ Engineering support,
  ■ Supplier identification,
  ■ Configuration updates of production test equipment.

7.3.2.3 Technical Documents

Technical manuals often do not match the production configuration of the equipment supported. Linking changes during the development effort to technical manual updates is often forgotten or neglected during other activities. The manuals can be unintelligible and not user-oriented (especially with software). The team should know the documentation requirements and the manuals should receive phasing and milestone checking like every other activity during the process. With the U.S. Department of Defense, this checking of the activity is called a "physical configuration audit," an important part of configuration control. Figure 7.4 shows a possible manual hierarchy for user and technical manuals.

7.3.2.4 Reducing the Risk

The following presents ideas for reducing the risk inherent in obsolete technical manuals:

■ Outline a clear delineation of customer and supplier responsibilities in the development, verification, validation, and publication of technical manuals in the project plan. Many times, the automotive customer will produce its own technical manuals.

■ Use automated processes (such as the use of computer-aided engineering drawings as illustrations) in technical manual preparation.

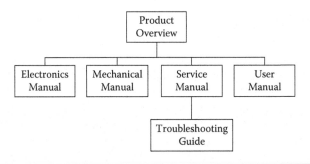

Figure 7.4   User manuals.

- The purchasing function analyzes technical options for portraying information including embedded and paperless delivery.
- Validate and verify drafts before final preparation and publication.
- Use automated readability analysis to verify that the level of the document matches the level specified (e.g., the Rudolf Flesch score for readability).

The development of technical manuals must be keyed to support training requirements, engineering development models, equipment evaluation, initial production units, and update programs.

7.3.2.5   Deviations

A deviation is a permission request from a supplier to a customer that allows an alternative part or process to substitute for the normal part/process for some specified period of time or number of pieces (see Figure 7.5). All deviations require customer approval if they affect form, fit, quality, or function of the component or subsystem. In general, it is a good idea to consult or notify customers of any change—to let them assist in the decision as to whether the change impacts form, fit, function, or quality.

A reaction plan is a tool useful for manufacturing, design, service processes and embedded development. If the team crafted the process control plans with appropriate information in the reaction plan column, it may not need a deviation to support the temporary modification since it should already have PPAP approval to proceed.

Deviations from the requirements stated on the drawings or specifications can arise for different reasons during manufacturing. In some cases, the deviations are such that it is impossible to use the parts for production; whereas, in other cases the parts can suffice. The production team should track all deviations and their status. Deviations should be reviewed for

		Internal Deviation			
LOGO	☐ Product ☐ Process ☐ System	Origination Date: _____ DEV# _____ Originator: _____			

Customer: _____	Platform: _____	Supplier: _____	
Customer P/N: _____	Our P/N: _____	Part Number: _____	
Part Description: _____		Part Description: _____	

Source of Issue:			
☐ Purchasing	☐ Materials	Permanent Change: ☐ Yes ☐ No Eng. Change (if Yes): _____	
☐ Supplier	☐ Customer	Validation Results: ☐ Yes ☐ No Attach results (if Yes)	
☐ Design/Engineering	☐ Other: _____	Customer Authorization: ☐ Yes ☐ No Attach Authorization (If Yes)	
☐ Manufacturing		Cost Impact: ☐ Yes ☐ No Quantity (pcs):	

Description of Deviation (include cost impact details if Yes):

Root Cause:

Corrective Action	**Date**	**Responsible (Name and Signature)**

Required	**Accept**	**Reject**	**Authorization**	**Comments:**
☐ Department 1	☐	☐		
☐ Department 2	☐	☐	_____	
☐ Department 3	☐	☐	_____	
☐ Department 4	☐	☐	_____	
☐ Department 5	☐	☐	_____	This authorization is required if rejected by the Quality Representative
☐ Department 6	☐	☐	_____	**Accept Reject**
☐ Department 7	☐	☐	_____	☐ ☐ _____
☐ Department 8	☐	☐	_____	This deviation will expire 60 <u>days</u> from the issue date.
				Expiration Date: _____
				Extension Date: _____

Figure 7.5   Deviation example.

1. The requirement/consequence class stated on the drawing (the degree of influence on the final product),
2. Possible complications to machining and assembly operations,
3. The time available for measures taken.

Appraisal results in one of the following measures:

1. Scrapping the part,
2. Adjusting or repairing the part,
3. Approving the part for a certain duration or for a certain quantity of parts,
4. Accepting the part and modifying the technical documentation (permanent acceptance of the deviation).

7.4   Cost

7.4.1   Delivery Performance

In many project contracts, there are penalties put on the supplier for late performance. Late deliveries that affect the customer's expected material receipts are more than an inconvenience. Even if there are no penalties contained within the contract, the customer's action to mitigate any missed performance will often be charged back to the supplier. Some examples of these noncontractual penalties are

1. Customer rework of part
2. Customer rework of installation (typically field work)
3. Unexpected handling steps forced upon the customer
4. Part acceptance and technical document modification.

7.4.2   Design for Manufacture

All design for manufacture (DFM) reviews should be complete by start of production. If the reviews are formal, the DFM team will release a document indicating the status of the process. In some cases, it will not be possible to eliminate all of the issues discovered during the DFM reviews, but it is necessary to manage the open issues to reduce risk. In many cases, the DFM team will rate the issues disclosed with some kind of system (for example, red, yellow, and green markings) to quickly reflect the status of the process.

7.4.3   Design for Assembly

Everything said about DFM applies to design for assembly. The fewer the steps in the assembly process, the quicker and easier the product can be produced. The team might also consider design for disassembly in the case where rework becomes necessary when the product is expected to be field-maintainable.

7.5   War Story

7.5.1   Last Minute Material Failure

The engineers designed a product and then determined that the liquid crystal diode (LCD) component exhibited abnormal behavior under high heat and humidity. The project team revealed this risk to the customer, who then decided to delay production until the supplier implemented a suitable corrective action.

Chapter 8

Failure Reporting, Analysis, and Corrective Action System (Phase)

8.1 Delivery

8.1.1 Importance of Failure Reporting, Analysis, and Corrective Action System

Quick identification when a failure happens means a quicker response to the problem and subsequent corrective action. This situation is no exception within the project environment. Whether in the product development, production support, or postlaunch customer effect, prompt determination of root cause and implementation of corrective action are critical. If the development team does not report and correct a problem in the development activity quickly, the corrective action delays production or causes nonconforming material to arrive at the customer's loading dock.

According to the Automotive Industry Action Group (AIAG), this phase starts at the beginning of the project and continues past the production start. This phase provides feedback for all of the preceding processes. That is, the failure reporting and corrective action system should function through the project and afterward.

In essence, the failure reporting, analysis, and corrective action system (FRACAS) provides the essence of a control system, with appropriate feedback and subsequent control provided to the overall development system. It is true that corrective action is reactive; however, the corrective actions become *preventive action* for future activities of the same genus.

We cannot overemphasize the importance of this part of the development/project system. Too many times, project managers will allow small problems to become large catastrophes due to the lack of alarm when an issue crosses a threshold. The FRACAS is so important that we believe enterprises should maintain it enterprisewide and not just for specific projects.

The concept of FRACAS is applicable to all variants of design (embedded or otherwise), service processes, and any other support function. That is why it is critical that the firm deploy this capability through all functions.

8.1.1.1 Initiate

At the beginning of the project, systems are to be in place that handle fault handling and corrective actions. This system will last during the project. Some versions of FRACAS will include change requests also. The project manager can make a decision to include risk management under this control system if his or her database tool supports this function.

8.1.1.2 Planning

A significant portion of the planning for FRACAS involves the setting of threshold values that will sound an alarm (see Figure 8.1). An example would be any deliverable that is overdue. Keep in mind that the project manager can create covert phantom due dates that *precede* the real due date to drive the system. If the deliverable has not met the real due date, it has already missed the phantom due date and, thus, the warning flag should be flying high.

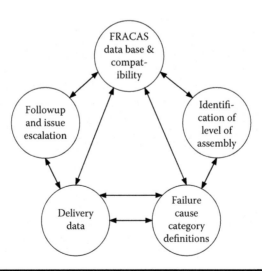

Figure 8.1   Elements of a good FRACAS.

One of the key activities of a project manager lies in the acquisition of product materials, especially during the development process. Most commonly, the enterprise will see a discontinuity between preproduction acquisition and launch and postproduction Manufacturing Resource Planning Software (MRP) type acquisition. The project manager can set thresholds for material by exploding the bill of materials and soliciting a long lead-time analysis from the acquisition organization. The project manager takes the results of this study and back schedules from the earliest need date (pilot runs or run-at-rate) and then adds some safety lead-time for variation in the quoted lead-time. By doing this, the project manager will end with the proper material at the optimal time of need.

During the early stages of production, the minimum order quantities should be set to a small value and permission given for outgoing air freight. Once the system settles down and the production team expects no further significant changes, the minimum order quantities can increase (achieving better economies-of-scale), lead-times may move out, and freight may only be by boat or land.

8.1.1.3 Execute

We can set time thresholds for corrective actions—violeting the threshold results in elevotion of the corrective action request to a higher level of management. The FRACAS needs its own thresholds; to whit, any corrective action that does not close within some predetermined duration should become elevated as a topic of concern. The quality engineers (or other designated individuals) should scan the FRACAS day-to-day and review it weekly.

When rigorously used, the FRACAS should reduce schedule/budget quality variation by pulling the project back within planned boundaries. The team can use a database tool that can trigger on due dates, sending a message (an alarm) to the project manager that the team has passed the threshold.

8.1.1.4 Control

As noted at the beginning of this section, the FRACAS serves as part or all of a project control system. Failures, changes, new risks, and other activities that lead to schedule/budget/quality variation should feed back into the database.

Significant to this portion of the FRACAS is the usual triad: cost, schedule, and quality. When a decision to *not* implement a corrective action occurs, the engineers should document it for future reference. Additionally, the decision should receive a risk assessment and the risk, if significant, should be tracked through the life of the project.

8.1.1.5 Closing

In a sense, the FRACAS never closes; it shifts ownership from the project manager to the production side of the house. When the handoff occurs, all corrective actions should be complete. The production facility should be able to use the same database to track issues through the life of the product (or, better, all products).

According to AIAG, the output from this phase is:[1]

1. Reduced variation,
2. Customer satisfaction,
3. Delivery and service,
4. Part acceptance and modified technical documentation.

8.2 Product Integrity and Reliability

8.2.1 Risk

8.2.1.1 Failure Reporting System

The ultimate objective of FRACAS is to implement corrective actions to prevent failure recurrence in product or process (see Figure 8.2). Issues occur when:

1. The flowdown of requirements from higher tiers to lower tiers is not systematic
2. Analysis of the failures has not been put forth as a requirement

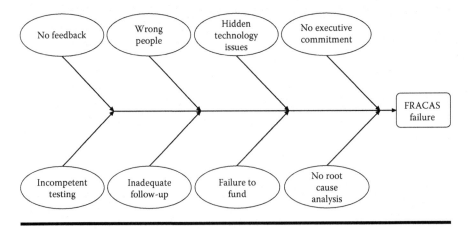

Figure 8.2 Typical FRACAS failings.

3. The prompt closeout of corrective actions has been overlooked
4. Notification to higher management about problem areas is missing

ISO/TS 16949 requires corrective action and the results of the corrective action constitute required records. Generally, suppliers and customers will retain their failure reports and corrective actions in a database and issue reports on a regular basis (e.g., weekly).

8.2.1.2   Reducing the Risk

The corrective action team should do the following:

■ Centralize implementation and monitoring.
■ Impose systematic, consistent requirements on all supply tiers and customer activities.
  ■ Report all failures.
  ■ Analyze all failures to sufficient depth to identify failure cause and necessary corrective actions. Automotive companies will use a formal problem-solving technique such as "Eight Disciplines" (8D) or Kepner-Tregoe in order to search for root causes.
  ■ Close all failure analysis reports within a well-defined interval after failure occurrence.
  ■ Alert corporate management automatically to duration problems.
  ■ Alert corporate management automatically to solution problems.
  ■ Arrange for lower tiers lacking facilities for in-depth failure analysis to be able to use laboratory facilities to conduct such analyses.
  ■ Prioritize criticality of failures consistent with their effect on product or process performance.

A FRACAS will be effective only if the reported failure data are accurate. The product support team will initiate the failure reporting system with the start of the test program and it continues through the early stages of development.

8.3   Cost

One high-quality, multiplant, enterprisewide software package is xFRACAS by Reliasoft. The package uses the Web browser as a "window" into the database and is accessible anywhere a Web browser is usable. An investment for five to six plant enterprises can run into tens of thousands of dollars. On the other hand, returned merchandise (zero miles) and field failures (warranty) can easily run into millions of dollars a year for such an organization. We can make the business case that the FRACAS actually saves us money.

8.4   War Story

8.4.1   Unplug It

There was a recurring failure mode in a product. After early investigation, the root cause remained hidden. At one point, a representative from the supplier commented to the supplier quality assurance (SQA) representative that all that had to be done to *bring* the component back into conformity was to unplug it. The SQA did not receive this information well since it was clearly a feeble attempt at humor with a bad product.

Chapter Notes

[1] Automotive Industry Action Group, Advanced Product Quality Planning and Control plan (APQP), (Southfield, MI, AIAG 1995) p29.

Chapter 9

Product Support

The goal of the product support phase is to make sure the product does indeed meet the expectations of the immediate customer and solve problems that can happen during the handoff and original equipment manufacturer's or customer's early production start. Support is often required during the transition from a development project to a steady state manufacturing activity. This is a transition from project activities to operational activities. This support starts at the supplier's manufacturing line and is often performed with product development staff. Often the product being supplied to the customer is a subsystem of a larger system. In these instances the customer may require some support, which takes the form of technical resources periodically available at the customer's facility. Even if technical support is not required by the customer, it is beneficial for the supplier to review the customer's handling of the product at the customer facilities. If off-site support is not required, many organizations build in product support from the development team as part of the launch process. This can be an arbitrarily assigned time such as 45 to 90 days after production start.

9.1   Delivery

9.1.1   Project Management Responsibility

The project manager is responsible for delivering a product through the process to production and project closure. This obligation requires support from the developers. Early in the project, the project manager's responsibility is to uncover the customer's expectations for product support during the launch phase. The project manager is not necessarily responsible for the action required to meet these expectations. However he or she is responsible for managing the customer's expectations. The project manager must ensure the organization meets the customer's expectation. The time to negotiate what to include in product support does not happen when there is no time to plan but only time to react. This is not to say that the team will see no additions or deletions in the project during the launch and product support phases. The project change management systems are used to alter the scope of the deliverables or project support. However, the team should identify the expectation, if feasible, early.

- Product documentation
- Part service literature
- Personnel and expertise
- Customer manufacturing facility support

9.1.2   Handing-Off the Project

Often, the launch program is a phased introduction with a start of fewer parts used than full volume followed by a monotonic increase in production volume (ramp-up) that relieves some of the risk. Ample opportunities for failures or perceived failures remain. Many times it comes down to the understanding of the product by the people on the production line. This suggests substantial training before launch could reduce some of these issues. Launch support is part of the effort to instill understanding of the component and how it merges with the rest of the system and with the people on the floor.

Figure 9.1 shows one example of closing activities for the various functions that exist in a design/manufacturing facility. As usual, our example uses the automotive documents. Keep in mind, these documents have general application regardless of the type of organization.

Launch is one of the most critical periods in a project. In many cases, it is the actualization of the project and also the time of highest risk with the least "wiggle room." At launch, the project manager must achieve a confluence of engineering, acquisition, materials management, shipping, accounting, production, and quality and bring the project in under budget, ahead of schedule, with no errors.

In the automotive world, launch means that the project has executed the following (at a minimum):

- Delivered a production part approval process (PPAP),
- Updated design and process FMEA documentation,
- Closed pending corrective actions,
- Verified and validated the product,
- Verified and validated the production processes,

Project Management	Action items–design review (should be closed) Pending changes Risks & contingencies Update project status table
Electrical Engineering	Design compliance
Mechanical Engineering	Design compliance
Embedded Software	Design compliance Verification of downstream analysis (customer process) Software release Issue firmware release notice
Printed Circuit Board Layout	
Labs	
Quality	Quality & yield compliance
Manufacturing- Surface Mount Solder	Manufacturing process compliance Control plan PFMEA
Manufacturing- Hand Assembly	Manufacturing process compliance Control plan PFMEA
Manufacturing- Final Assembly	Manufacturing process compliance Control plan PFMEA
Production Test Equipment	Manufacturing test compliance
Accounting	
Materials	
Information Technology	Manufacturing test compliance
Strategic Purch.	Preferred Parts List compliance Standard cost compliance
Customer service	Production plan & forecast compliance
Customer	

Figure 9.1   Project handoff.

- Executed pilot runs and run-at-rate,
- Secured permission to proceed from the customer,
- Secured permission to proceed from any other tiers, if present.

These concepts generalize for service processes also. Reviews and critiques of the proposed services and processes are just as valid as hardware or other physical part reviews (DFMEA and PFMEA). Verification and validation activities take the form of target customer interviews and surveys. Pilot runs are analogous to phase introduction of the service into different geographic locations, with closure of action item lists of problems found driving improvements.

9.2   What Happens during Launch

Organizations should expect a modicum of chaos to occur during the launch activity. If we are speaking about manufacturing, the materials ordering system will not have reached a stable point where the standard gross margin is asymptotic to the expected value. Minimum order quantities should be in the vicinity of *one* to eliminate lifetime supplies of parts the customer rarely orders.

Figure 9.2 shows what a company who desires a successful launch might expect to see. Initial profit is negative; in other words, the organization loses money on each delivery to the customer. On the other hand, the customer is receiving final assemblies promptly. During this launch period, which in the automotive world can often take a few weeks to a couple of months, the supplier accumulates enough information to be able to begin to adjust the materials' supply and shipping systems, moving the standard gross margin in the direction of profitability. Often, the customer will begin to seek engineering changes as the new design makes more

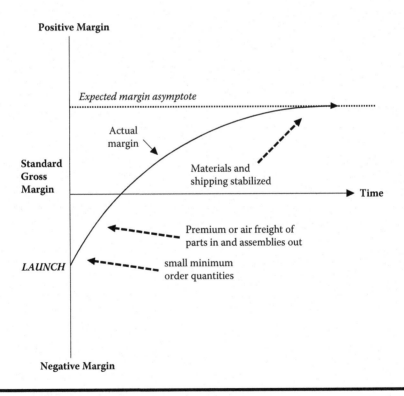

Figure 9.2   Realistic launch expectations.

frequent contact with real usage. The launch team (or production support team) must evaluate each change for time and cost, sending prompt quotes to the customer. Additionally, the engineering changes insert elements of risk as the configuration of the target design now becomes less stable. We expect the same kind of effect to take place during the implementation of a service; that is, the initial period is chaotic as the service team begins to deal with real customers and starts making adjustments for service delivery and quantity.

By the time of the launch, the embedded designers should have a stable release; however, they should be alert to probable changes to the software as the customer realizes their specification did not adequately describe what the end customer wants to see. The embedded developers will see the same kind of "churning" that the hardware developers can expect to experience.

9.2.1   Crashing Product Support

The project support phase objectives exist in order to

- Ensure that the customer uses or handles the product as prescribed by the supplier,
- Assist in disseminating information design performance with the customer production staff,
- Identify early problems.

Crashing product support, either by shortening the duration or by insufficient presence at customer locations, can have a damaging effect on the opinion of the supplier. The situation where postlaunch support is minimal will be particularly damning if the project had a difficult execution and, in the end, the supplier does not provide on-site support.

9.3   Product Integrity and Reliability

9.3.1   Launch Containment

The customer should return failed parts promptly to allow for calculation of the severity frequency of any bad material and to begin ascertaining the reasons. We have witnessed launches where this early production support was nonexistent or inadequate and early failed material was not promptly shipped back to the supplier. Delay in returning material is a delay in determining causes and corrective actions, leading to increasing amounts of material making it through production and into the field. Many customers require the supplier to have relevant expertise available at the customer site during product launch, usually consisting of either a resident engineer or temporary staff. On-site staff expedites material return. The failed material can be reviewed by both the supplier and the customer. This allows for opportunities for a common understanding of the problem.

9.4 Cost

9.4.1 Extensibility

With software, extensibility means it is able to have additional features added without degrading the core code. Designing a system that is built to have additions means that the product has excess capacity, often in the form of processor speed or random access memory. This excess capacity comes at an additional cost. In general, the embedded development is wise to plan for both excess speed and memory to allow for probable design changes from the customer.

9.4.2 Emergency Shipment

At the start, the supplier is unlikely to be recouping much profit from the product. Late shipments or emergency "air drops" to keep the customer's delivery schedule negatively affects the payback period on the project. In the automotive environment, delivery failures receive fines from the customer, sometimes at the level of millions of dollars per day of shutdown.

One approach that can reduce the probability of late deliveries would be the use of a process FMEA on the project itself. The project team can analyze each step in the plan and devise responses to negative events long before they happen. The best use of a project FMEA occurs when the planning is so robust the *wicked* event never happens.

9.5 War Story

9.5.1 Supplier/Customer Relationship

A product had been collaboratively developed by the customer and the supplier with the customer supplying specifications for the product. This was a custom product and not one the supplier would be able to sell to another customer. The customer committed to a certain number of years of production at a defined volume of parts per year.

Later, after an organizational acquisition, the customer decided to develop a new similar product to take advantage of the higher volumes and synergies with the new organization. This decision meant an earlier than expected termination of the existing production run. The customer communicated that the production run would be a few years shorter than expected. The supplier, in response to this shortened production run, raised the price of the component to recoup the production line setup costs. This created much friction between the supplier and the customer thereby causing damage to the relationship.

In the automotive development world, a supplier will apply to the customer for *cancelation charges* based on the contractual agreement. We see no reason why embedded development groups and service businesses can not write cancelation charges into their contracts.

9.5.2   Field Failure

The description that follows is a composite of a number of stories that essentially are the same.

It is often not possible to verify all combinations or permutations of a product or system. For very complex systems or systems that exist in multiple incarnations or variants, the combinations can be so high that there is no chance to perform all of the required verification. This can be especially true of heavy vehicle applications, particularly since there are often numerous variations as well as aftermarket systems sold for these vehicles. We have a number of experiences in which a variety of subsystems are put on the vehicle. Occasionally, these systems interact in an unpredictable way.

One particular combination produced an unpredictable response. Upon much investigation, it was possible to determine the combination of stimuli acting upon the control module to generate the failure. This sequence of events was described to the supplier. The supplier reviewed its software with this understanding and found an error in the code that would allow the problem to exist. The supplier was able to pinpoint the software malfunction rather quickly and make the necessary alterations to the software and reverify. It is possible to conceive that if the supplier had a rigorous code review, with a critical eye, then the failure could have been found prior to production start and saved the supplier and his or her customer the required update of numerous vehicles.

Chapter 10

Program Management

10.1 Program Management versus Project Management

Some people use the terms "project management" and "program management" interchangeably. In automotive and government development environments, these names do not refer to the same actions. The skills required to manage a program are analogous to those for a project.

A *program* is a collection of projects, Figure 10.1, handled concurrently for either efficiency or due to some synchronization constraint such as a vehicle launch or, for that matter, any final product or service composed of multiple "threads." An example of a project is the development of the suspension system for a vehicle. Delivery of this suspension system would mean a successful project. The program is the collection of projects that must complete synchronously for the launch of the vehicle to be successful. This synchronization requires coordination of the development of multiple subsystems.

The program manager sets the master schedule and all other project components must support this master schedule or the launch of the vehicle is at risk. With U.S. government projects, this schedule is called the "integrated master schedule" and the plan is called the "integrated master plan."

The program manager manages each project manager and his associated project and is responsible to ensure that these supporting projects lead to the success of the overall project or program. In essence, the program team will functionally decompose a large program into more manageable projects. Hence, the program manager performs the function of reconciling the schedules and deliverables of each project.

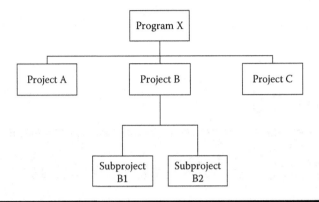

Figure 10.1 Program management and project management.

10.2 Program Manager Skill Set

Since the program manager must understand the subsidiary project, it is clear that a program manager's skills would be similar to those of a project manager. However, this person will likely be called on to mentor and help solve higher-order problems than those capable of resolution by project managers. Ideally, the program manager would be an individual who has demonstrated exceptional project management and diplomatic skills, thus making him or her an excellent candidate.

Chapter 11

Final Thoughts

11.1 How Many Programs Work Concurrently?

The number of concurrent programs that an enterprise can handle can only be determined empirically. In general, the program office should look for candidates with demonstrated experience at managing multiple projects. If multiple projects compose a program, then the enterprise would definitely seek an experienced program manager who can balance stakeholder needs, organizational needs, and human resource needs. The use of contractors helps adjustments to meet peak demands above the firm's steady state, but will only partially solve problems. The tradeoff to contracting projects is loss of control and failure to retain expertise.

11.2 Generalizing Automotive Techniques to Project Management

Any place where a team develops a product, the automotive processes and documentation are applicable. Let's review some examples:

- Production part approval process (PPAP) applies to software development
- Risk assessment and management applies to the development of services, software, and hardware
- Failure mode and effect analyses apply to any activity or object where we desire to anticipate and manage problems before they happen
- Measure system analyses apply to nearly any activity, from public school systems to heavy manufacturing
- Statistical process control can be used in any situation where we would like to separate assignable events from random variation

- Documented launch processes are usable by service providers, software developers, and hardware developers.

The point here is that the automotive is disciplined but not constrained to automotive development only.

11.3 Future Trends

Existing trends growing into the future will continue to increase in importance.

11.3.1 Globalization

What is it? Globalization significantly affects the execution of projects due, in part, to the difficulty of communicating across myriad time zones. We have both worked on teams of projects that are distributed around the world: software development in Europe, hardware design in North America, testing in Latin America, component supply from China and India, and manufacture in the Baltic states. We work at companies who are global in composition; however, with the scarcity of expertise and costs driving projects, we anticipate this distributed development approach gaining ground.

Devising ways to facilitate teamwork in these distributed teams is a challenge. We have witnessed instances where a small, dedicated, and talented project team, when colocated, can produce extraordinary results—an effect that often derives from the immediate and face-to-face communication within the colocated team. Getting distributed teams to perform to the same level may be difficult or impossible. Microsoft provides NetMeeting,® which allows a computer user to demonstrate a program in real-time with viewers able to contribute directly.

Internet Collaboration tools that are Web-based or that are easily accessible can help solve some of the challenges associated with distributed development. There are a number of tools that allow access to project information via the Internet. These tools are developed specifically for project management of distributed teams. However, there are other software applications that we can adapt to improve the project ability to deliver; for example, Microsoft supplies the Sharepoint® tool to support distributed work-groups. Blogs are a method of providing a journal for the project activities and a forum for distributing project information and challenges to the rest of the team. Bulletin boards such as *Vbulletin*® can tie a distributed team together with essential information and progress. A distributed project team might also consider the implementation of a wiki—special software that allows users to create, edit, and link Web pages themselves.

If a distributed team is able to function effectively, reward and recognition may also become problematic. The same issues of location and time apply. One tool that can give the illusion of colocation is video conferencing.

11.3.2 Reduction in Force

Staff reduction or downsizing is not a new trend. Force reductions can cause havoc with project plans because team members may no longer be employed. Force reductions often occur during periods of poor financial performance. Diminishing the work force provides an instant boost to operating profits. However, the force reduction also represents a reduction in tribal knowledge as that knowledge leaves the enterprise. For a project to be successful, it must have the necessary competencies needed at the appropriate time and there must be some motivating factors to meet the objectives. Therefore, it is obligatory for management to negotiate the separation of team members in such a way that the program manager can compensate for or replace the missing individual.

11.4　Troubled Project

An example of an excellent book on the subject of troubleshooting projects and determining corrective actions is by Boris Hornjack, called *The Project Surgeon.*

It does not take much effort to find the symptoms of a failing or failed project—all you really have to do is talk to the participants and find and review data. The problem often is that there are no key performance data collected during the project to perform the assessment. It is necessary to identify possible project failures early on and devise data collection and measurements to reveal negatives quickly so they can be managed.

The same technique used by engineers to debug designs and manufacturing lines are also applicable to project management. Illustrated in Figure 11.1 is an Ishikawa or fishbone diagram to identify root cause.

In order to troubleshoot projects, you must have experience, intuition, desire, and, above all, *data.* Figure 11.2 is a graphic of a chart that can

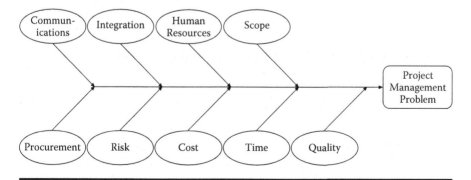

Figure 11.1　Project management use of Ishikawa diagram.

Project Number

Activity	Status				Q	D	C	F	TOTAL
	Yes	No	Late	Inadequate					
	100	0	50	30					
Program Definition									
Quality Function Deployment Activities	1			1	75			25	70
Preliminary Engineering Bill of Materials	1				25			75	100
Product Specifications	1						25	75	100
Specification Reviews	1			1	75			25	70
Establish Quality and Reliability Targets	1		1		100				50
Product Assurance Plan	1				100				100
Preliminary manufacturing Bill of Materials	1					50		50	100
Preliminary Process Flow Diagram	1				25	75			100
Special Process Characteristics	1			1	75	25			70
Software Quality Assurance Plan	1				75			25	100
QDCF Sum For Next Gate									86
Product Development									
DFMEA	1				75			25	100
DFMA	1		1		50	25	25		50
Key Product Characteristics Identified	1				50			50	100
Design Verification Testing	1			1	75			25	70
New Equipment and Tooling Requirements	1				50	50			100
Gauges R&R	1				50				100
Product Test Equipment	1		1		75			25	50
Engineering Bill of Materials Released	1				75			25	100
QDCF Sum For Next Gate									83.75
Process Development									
PFMEA	1			1	75	25			70
Key Control Characteristics	1				100				100
Process Control Plan	1		1		75			25	50
Special Process Characteristics	1								100
Process Flow	1				50	50			100
Process Floor Plan	1				50		50		100
Pre-Launch Control Plan	1				75	25			100
Process Instructions	1			1	100				70
Process Verification	1			1	75			25	70
Product and Process Quality System Review	1		1		100				50
Measurement Systems Analysis	1								100
Packaging Specification	1		1		50				50
Packaging Specification Review	1				75	25			100
Process Capability Study	1		1		100				50
EDI	1						25	75	100
QDCF Sum For Next Gate									81
Validation of Product and Process									
Design Validation Plan and Report (DVP & R)	1			1	75			25	70
Preliminary Process Capability	1							100	100
Bench Testing		1						100	0
Systems Testing	1							100	100
Measurement Systems Evaluation	1		1		100				50
Production Part Approval	1							100	100
Packaging Evaluation	1							100	100
Production Control Plan	1				50	50			100
QDCF Sum For Next Gate									78
Release to Production									
Process Sign Off	1			1	100				70
Trial Production Run	1					50		50	100
Pilot Runs	1		1		100				50
Run at Rate	1				50	50			100
Production Test Equipment Evaluation	1				75	25			100
Design Change Notification	1		1		50			50	50
QDCF Sum For Next Gate									78
Weighted Value									81.25

Figure 11.2 Product development success.

be used to predict the product development output quality. This sheet was created in Excel®, where the left column represents the list of tasks deemed necessary to secure the product quality, function and deployment, and the project cost. The "Status" column is an assessment of the suitability or efficacy of the line item.

For example, the manager decides the project design review is necessary; however, he records it as late or inadequate. This has some possible negative effect on the successful delivery of the product unless the manager made a prior assessment of the consequences of failure or of insufficient design review (see Figure 11.2).

11.5   Project Manager Origins

11.5.1   *"The Accidental Project Manager"*

In the past, project managers often came from any part of the organization (or from outside the organization) and received little formal training in project management. With increased expectations and importance put on this area in terms of delivering for the organization, this tactic is becoming too risky. Many organizations are moving to increase the knowledge and credibility of project managers by requiring training and pushing individuals to meet increased expectations. In reality, this trend should have been the norm all along. It is the exceptional individual, for example, who for example can be an electronics engineer without the required training and experience. The position of project manager is no less critical for organizational success than competent, trained, experienced, and committed project managers. Placing an inexperienced individual into a project position does not ensure the project's success nor will it boost the individual's confidence in being able to meet the demands of the position.

The U.S. government was way ahead of corporations with regard to the training and education of project managers. The government courses are 12-weeks long and extremely rigorous. The development of multibillion dollar weapons systems is serious business!

11.5.2   *Technical Talent as Project Managers*

We have been in organizations or have worked with organizations that believe strong technical talent transfers easily to a good or great project manger. There is little doubt that technical knowledge helps ensure that the technical details receive adequate attention. However, people with strong technical talent can be distracted by the desire to participate in the technical aspects of the project instead of *managing* the project. Additionally, the human resource management skills may not be as refined in technical people. This is not a stereotypical comment about the engineer who does not possess people skills; it is a statement that originates in the knowledge that a person writing code on a PC does not require as much interaction with people as does a person who must resolve conflicts with priorities and other constraints. The one thing that can be said is "strong technical people are strong technical people." They may not be especially good candidates for a project management position, particularly if they are unprepared for the transition by lack of formal training or mentoring. The project manager must know when he or she *must* use any technical capability he or she may have, and when he or she *could* use that ability. We have been in heated meetings where the new project manager (software lead engineer) exploded at his staff over a particular design solution with the customer present.

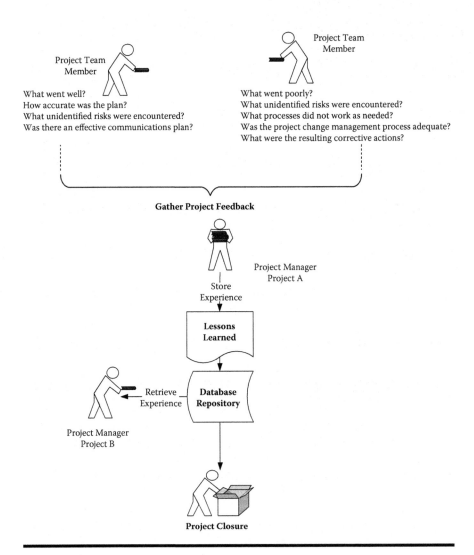

Figure 11.3 Structure of lessons learned.

11.6   Outsourcing Project Management

Although outsourcing is not new, we do not know of a particular trend to outsource project management responsibilities. As long as it is possible to fulfill an organizational need in a competitive way, we expect to see outsourcing. A project manager must be able to span the various departmental organizations within the corporation. This people network would seem a rare thing for a contracted person (person not within the core of the organization over time) to have at his or her disposal. When a product is developed entirely by an outside organization, the PM is typically a member of that developing organization. There is a counterpart project manager at the ordering customer. Since he or she is responsible for making connections to various other departments of the organization, even this person would rely on a network of people to be able to successfully deliver the project.

11.7   Importance of Lessons Learned

In the early period of an organization's involvement with project management, it may not have much historical project information. For an organization to adequately meet future demands, it should make use of lessons learned documentation; that is, if the enterprise uses after-action reviews to produce these documents. Reviewing lessons learned at a project's close provides the opportunity to assess schedule, cost, quality, and other pertinent project metrics (see Figure 11.3). This information should be used to advance the enterprise capabilities by allowing an adaptation of organizational processes and to improve the operating environment of the organization. When done well, the team achieves a better understanding of organizational weaknesses and instigates aggressive corrective action.

The figure shows a potential structure for a lessons learned submission. Often, the most difficult part of lessons learned activities is the creation of a database or a wiki that allows for rapid, easy retrieval of the desired information. If using a database, it is essential that the originator use language that allows for searching.

Appendix A

The 18 PPAP Documents

All production part approval process (PPAP) submissions include a part submission warrant (PSW), which tells the customer basic information about the submission package. The Automotive Industry Action Group (AIAG) defines the content of the PPAP as:[1]

1. Part submission warrant (PSW)
2. Design records/drawings
3. Engineering change documents
4. Design failure mode and effects analysis (DFMEA)
5. Process flow diagram
6. Process failure mode and effects analysis (PFMEA)
7. Dimensional results
8. Material/performance test results
9. Initial process study
10. Measurement system assessment (MSA) studies
11. Qualified laboratory documentation
12. Prelaunch control plan
13. Process control plan
14. Appearance approval report
15. Bulk material checklist
16. Product sample
17. Master sample
18. Checking aids
19. Customer specific requirements

The PPAP process and documentation attempts to provide evidence that the supplier understands all of the customer's requirements, and that the development work and the process work are capable of consistently achieving those requirements under production rate stresses. Any change

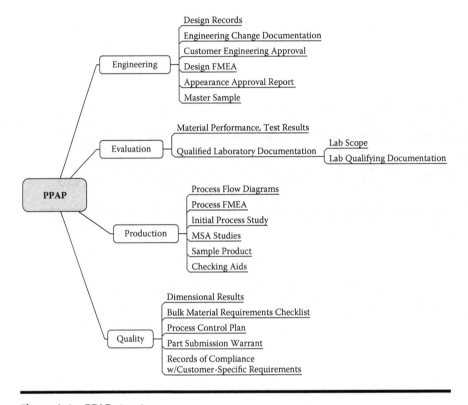

Figure A.1   PPAP structure.

to the product or processes must be with the customer's knowledge and consent. These changes will typically require a resubmission of the PPAP documents (see Figure A.1). Depending on the nature of the change, the resubmission could be from warrant only to all PPAP activities including product samples, complete rework of the documentation, and review of the supplying organization's manufacturing facility.

Chapter Notes

[1]Automotive Industry Action Group, Production Part Approval Process (PPAP), (Southfield, MI, AIAG 2006) p18.

Appendix B

System Requirements Review (SRR)

Not all reviews are formal activities. In many cases informal review activities are used between engineers to improve the quality of the resulting requirements document. All subsequent discussions of reviews are from this formal perspective (see Figure B.1); however, in our experience, these informal reviews are equally important to securing the quality of the requirements documentation.

System requirement reviews ensure the requirements are well identified, tangible, and meet the required system performance expectations. The reviews also assess the probability of the proposed system meeting the cost and schedule targets. Figure B.1 illustrates the placement of reviews within the process.[1] The early identification of the risks highlighted during the technical review allows for actions to "short circuit" the risks.

The reviews are periodic through the generation of the system requirement (development phase). These reviews are conducted to establish direction and assess progress of the system requirements effort. Postspecification development reviews are frequently performed for a supplier's edification.

Topics for review can include:

1. Requirements verbiage
2. Mission and requirement analysis
3. Functional flow analysis
4. Preliminary requirements analysis
5. System/cost effectiveness analysis
6. Trade studies
7. Synthesis
8. Logistics support analysis

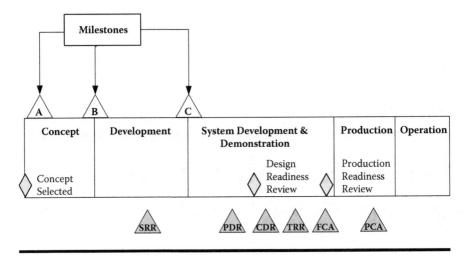

Figure B.1 Technical review distribution.

9. Speciality discipline studies
10. System interface studies
11. Generation of specifications
12. Program risk analysis
13. Integrated test planning
14. Producibility analysis plans
15. Technical performance measurement planning
16. Engineering integration
17. Data management plans
18. Configuration management plans
19. System safety
20. Human factors analysis
21. Value engineering studies
22. Life-cycle cost analysis
23. Preliminary manufacturing plans
24. Manpower requirements/personnel analysis
25. Milestone schedules

General rules that apply to all reviews apply to all subsequent appendices regarding reviews. A review is only effective if adequate preparation time has been allowed. Those performing the critique must have analyzed the material before the meeting. A way to ensure that all material is reviewed before meeting is to ask those who attend the first meeting for their documents as the "entry ticket" to the discussion group.

The reviewing staff must be subject matter experts and should answer from a critical technical perspective. Internal company politics should be quiescent during the review. Political sensitivities may temper the ultimate decisions made from the review—political requirements may end up as product or process requirements.

There should be a meeting facilitator to keep the conversations under control as well as a meeting scribe. It is possible to update the various documentation on the spot with the feedback from the participants. An overhead projector, laptop computers, and Netmeeting® allow team members to see the material under review and the changes being made in real time.

Chapter Notes

[1]Defence Acquisition University Press, Test and Evaluation Management Guide E5, (Fort Belvoir, VA, DAU; Jan 2005) p8–3.

Appendix C

System Design Review (SDR)

System engineering takes on different meanings for different organizations. Some organizations treat systems engineering as the technical areas of project management. In these organizations, the project manager focuses on the budget and scheduling. Afterward, a senior technical person handles all other technical and leadership responsibilities needed to deliver the product. This individual will manage the system that becomes the project. We will discuss systems engineering from the perspective of MIL-STD-499B. Systems engineering is a level of abstraction above the detailed development effort.

We conduct these reviews to evaluate the proposed system design. This review occurs when system characteristics are defined (see Figure B.1), including specific software items as well as hardware items. The system development work must achieve the level of detail the team believes will meet customer functional and performance targets. This review assesses the level of optimization, correlation to customer targets, as well as level of completeness of the system as defined. We often see iterations of these reviews, where the product for the first review becomes the *system baseline.* Each function should be traceable to customer requirements. This review can be a paper, model, or nonproduction prototype critique.

The SDR includes a review of the following items, when appropriate:

1. Systems engineering management activities, for example:
   a. Mission and requirements analysis
   b. Functional analysis
   c. Requirements allocation
   d. System/cost effectiveness

    e. Synthesis

    f. Survivability/vulnerability

    g. Reliability/maintainability/availability (r/m/a)

    h. Electromagnetic compatibility

    i. Logistic support analysis (i.e., shipping, service, maintenance)

    j. System safety (emphasis shall be placed on system hazard analysis and identification of safety test requirements)

    k. Security

    l. Human factors

    m. Transportability (including packaging and handling)

    n. System mass properties

    o. Standardization

    p. Value engineering

    q. System growth capability

    r. Program risk analysis

    s. Technical performance measurement planning

    t. Producibility analysis and manufacturing

    u. Life cycle cost/design to cost goals

    v. Quality assurance program

    w. Environmental conditions (temperature, vibration, shock, humidity, etc.)

    x. Training and training support

    y. Milestone schedules

    z. Software development procedures [Software Development Plan (SDP), Software Test Plan (STP), and other identified plans, etc.]

2. Results of significant tradeoff analyses, for example:

    a. Sensitivity of selected mission requirements versus realistic performance parameters and cost estimates

    b. Operations design versus maintenance design, including support equipment effects

    c. System centralization versus decentralization

    d. Automated versus manual operation

    e. Reliability/maintainability/availability

    f. Commercially available items versus new developments

    g. National stock number (NSN) items versus new development

    h. Testability

    i. Size and weight

    j. Performance/logistics

    k. Life cycle cost reduction for different computer programming languages

    l. Functional allocation between hardware, software, firmware, and personnel/procedures

    m. Life cycle cost/system performance trade studies to include sensitivity of performance parameters to cost

     n. Sensitivity of performance parameters versus cost

     o. Cost versus performance

     p. Design versus manufacturing consideration

     q. Manufacture in-house versus outsourcing ("make versus buy")

     r. Software development schedule

     s. On-equipment versus off-equipment maintenance tasks, including support equipment effects

     t. Common versus peculiar support equipment (usually one-off equipment)

3. Updated design requirements for operations/maintenance functions and items,

4. Updated requirements for manufacturing methods and processes,

5. Updated operations/maintenance requirements for facilities,

6. Updated requirements for operations/maintenance personnel and training,

7. Specific actions to be performed include evaluations of

     a. System design feasibility and cost effectiveness

     b. Capability of the selected system to meet requirements of the lower-level specifications

     c. Allocations of system requirements to configuration items

     d. Use of commercially available and standard parts (off-the-shelf when possible)

     e. Allocated inter- and intrasystem interface requirements

     f. Size, weight, and configuration that allow for economical and effective transportation, packaging, and handling

     g. High-risk long lead-time items

     h. The ability of inventory items to meet overall system requirements

     i. Value engineering studies

As with all reviews, positive impact on the project is contingent on the amount of planning and appropriate resources allocated and followup actions. The conclusion of the review should produce:

1. Action item lists,

2. Review minutes,

3. Log of identified risks,

4. Preliminary identification of production needs (key product characteristics),

5. The next update to the system baseline.

Appendix D

Software Specification Review (SSR)

The software specification reviews (SSRs) are tools to ensure the documentation supports the customer needs. These reviews are the input into the detailed development activities. If this input is poor, then the output will also be poor and will not fulfill ambiguous or missing requirements. If the product meets the customer's expectation, it will be by accident and not by deliberate action. The inability to meet customer requirements means many change requests, driving the project over budget, with risk for delay. The worst case would be if the product ships to the customer as scheduled, only for him or her to find out it does not meet expectations, usually after many field failures occur. Figure D.1 demonstrates where the review process fits into the specification process, and how the review results improve the specification.

Additional details and considerations for conducting reviews can be found in Appendix B and Appendix F. Without attention to detail and lacking due diligence, reviews quickly lose the ability to add to the quality of the finished product.

Reviews are a frequent, scheduled activity used as a tool by the project manager to control the progress of the project or program. Periodic reviews during the specification creation ensure the specification does not drift from the target. Often these reviews are not part of the formal reviewing structure and consist of smaller groups (specification author and other engineers, especially software engineers) in a much less formal setting. These reviews typically do not cover the entire scope listed below but address areas in advance of the formal review.

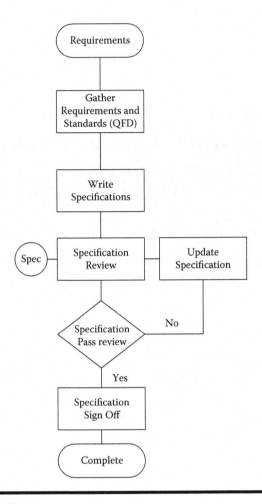

Figure D.1 Specification process.

1. Functional review of the software modules including:
   a. Inputs
   b. Processing
   c. Outputs
2. Software performance targets (execution time)
3. Data and control flow of the software modules
4. Internal and external interface requirements
5. Verification requirements for all software modules and interfaces
6. Software quality factors
   a. Reliability
   b. Efficiency

    c. Useability (and reuseability)
    d. Portability
    e. Flexibility
    f. Interoperability
    g. Testability
7. Updates from last review
8. Milestone review

Appendix E

Software Quality Assurance Plan (SQAP)

A software quality assurance plan (SQAP) defines the tactics, measurements, procedures, processes, and standards that the project management team believes necessary to deliver a quality product. The document itself should convey to the participants the agreed on methods, metrics, and controls to deliver the product. Periodic monitoring of the software and the project are then measured to the plan.

The SQAP is not a one-shot documentation effort. It requires constant review and provides comparison for the actual project execution and software development. If correctly performed, the SQAP will:

1. Define software development, evaluation, and acceptance standards to which the software development will be compared over time.
2. Illustrate results of these periodic comparisons of actual to planned, and will be delivered to management to keep it apprised of the project and software status
3. Reinforce and ensure the acceptance criteria for the product and project will be met

The Institute of Electrical and Electronics Engineers (IEEE) standards for software engineering are not as consistent as those from MIL-STD-498; that is, they have structural variances depending on which committee put them together.

IEEE-730 defines elements of the SQAP as:

1. Purpose
2. Reference documents

3. Management
   a. Organizational structure for SW development
   b. SW life cycle covered and specific tasks to quality assure the software
   c. Responsible parties for each task
   d. Relationship between tasks and project checkpoints
4. Documentation
   a. Software requirements specification
   b. Software design description
   c. Software verification and validation plan
   d. Software verification and validation report
   e. User documentation
   f. Software configuration management plan
5. Standards, practices, conventions, and metrics
   a. Identify specific standards, practices, and metrics
   b. State how compliance will be monitored and assured
6. Reviews and audits
   a. Software requirements review
   b. Preliminary design review
   c. Critical design review
   d. Software verification and validation plan review
   e. Functional audit
   f. Physical audit
   g. In-process audit
   h. Managerial reviews
   i. Software configuration management plan review
   j. Postmortem review (lessons learned)
7. Test
8. Problem reporting and corrective actions
9. Tools, techniques, and methodology
10. Code control
11. Media control
    a. Identify documentation required to be stored including copy and restore process
    b. Identify processes for protection of physical media from access and damage
12. Supplier control
13. Records collection, maintenance, and retention
14. Training
15. Risk management

Appendix F

Software and Hardware Reviews

IEEE Standard for Software Reviews 1028 provides definitions and uniform requirements for review and audit processes. The principles and tactics defined for the software realm work just as well for the hardware aspects of the embedded development work.

F.1 Overview

There are five types of reviews,[1] each one designed to meet an aspect of the software delivery and stakeholder expectations. In general, the goal of a review is to ensure the area under review receives appropriate critique. All of the reviews in these appendices are part of the overall quality system. Collaboration and communication improves output, identifies areas of risk, or reviews current status compared to the ideal. The five types of reviews are as follows:

1. Management reviews
2. Technical reviews
3. Inspections
4. Walk-throughs
5. Audits

F.1.1 Management Reviews

These reviews provide a formal and structured assessment of various documents and processes employed in software acquisition, development, delivery, and maintenance. The reviews monitor progress and effectiveness in meeting the needs of the organization, the product, and the project. Management reviews influence and support decisions made about resources, requirements, corrective actions, as well as scope changes to project and product.

Generally, management reviews fit into one of the categories below:

1. Monitoring progress of project
2. Determining the status of plans and schedules
3. Confirming requirements and their system allocation
4. Evaluating the effectiveness of defined management approaches (or impact of missing management) to
   a. Risk mitigation
   b. Resource allocation
   c. Corrective actions strategies
   d. Suppliers
   e. Customers
5. Assessing project scope fulfillment

All reviews follow a general process. Management reviews follow an outline similar to the following:

1. Identifying need for review (when)
2. Determining scope of review
3. Planning the review
   a. Structure (agenda)
   b. Participants (competencies) and roles
   c. Procedures
4. Preparing for meeting—connecting scope of review and documentation required
5. Reviewing
   a. Conduct meeting
   b. Identify specific areas of risks, corrections, or problems
   c. Assign actions to solving issues
6. Following up on actions generated
7. Assessing review results

Examples of specific areas of management reviews are:[2]

1. Anomaly (erratic performance) reports
2. Audit reports
3. Backup and recovery plans
4. Contingency plans
5. Customer or user representative complaints
6. Disaster plans

7. Hardware performance plans
8. Installation plans
9. Maintenance plans
10. Acquisition and contracting methods
11. Progress reports
12. Risk management plans
13. Software configuration management plans
14. Software quality assurance plans
15. Software safety plans
16. Software verification and validation plans
17. Technical review reports
18. Software product analysis
19. Verification and validation reports

F.1.2 Technical Reviews

The technical review is comprised of experts who evaluate software; however, these types of reviews need not be restricted to software. Hardware reviews are beneficial as well.

Typical technical reviews are

1. Software requirements specifications
2. Software design description
3. Software test documentation
4. Software user documentation
5. Customer or user representative complaints
6. Product hardware specifications
7. Maintenance manuals
8. Acquisition and contracting methods
9. Progress reports
10. Risk management plans
11. System build procedures
12. Installation procedures
13. Release notes

F.1.3 Inspections

The purpose of any inspection is to identify nonconformity or poor quality in the product and product compliance with specifications and standards. Inspections are similar to the technical review in that they are peer reviews and require a high level of competence. The team is typically four to five people, and typically takes 1 to 3 hours of up-front review and 1 to 3 hours of meeting review time.

Software inspections are not free form but, rather, they are rigidly structured. Properly conducted inspections improve:

1. Extensibility (maintainability)
2. Reliability
3. Quality
4. Team knowledge of the software distribution

These reviews reveal the software's level of refinement. Organizations often have style guides for writing code. Additionally, the level of completeness and correctness are assessed by directly tracing requirements back to the software modules and functions. Organizations that employ the use of inspections know that relying on testing for finding problems costs time and money and there is still no guarantee that the testing will find all problems.

The steps for inspection are

1. Identify inspection team members and roles,
2. Distribute review material,
3. Identify goals of inspection,
4. Review meeting,
5. Categorize and track defects found,
6. Develop action plan for corrective actions,
7. Schedule second review to assess corrective actions.

IEEE style software inspections can include:[2]

1. Software requirements specification (SRS)
2. Software design description (SDD)
3. Software test documentation (test plans and test descriptions)
4. Software user documentation
5. Source code
6. Maintenance manual
7. Troubleshooting and diagnostics manual
8. System build procedures
9. Installation procedures
10. Release notes

It is possible to extend this method further into the embedded hardware world by using the technique for:

1. Hardware or system architectures
2. System specifications
3. Hardware specifications
4. Hardware test plans
5. Hardware test descriptions
6. Hardware user documentation

Extended to the other activity types:

1. Process documentation
2. Management plans
3. Contract reviews
4. Material deliveries
5. Risk plans
6. Communications plans
7. Resource matrix
8. Requirements and constraints evaluations
9. Pugh matrix
10. Value engineering activities

F.1.4 Walk-Throughs

The purpose of the walk-through is to improve the software and develop strategies for meeting the software requirements and user expectations. Walk-throughs are used to refine the software implementation. Walk-throughs are much less rigorous and produce fewer results for roughly the same expenditure of time. The walk-through meeting is run by the document author. Walk-throughs are not typically reviewed by technical experts, but are often used in reviewing specifications to ensure all have an understanding of the document under scrutiny.

As in any good review, identifying the stakeholders of the document and making sure they are present is a must. Each reviewer in the meeting is provided with material to review before the walk-through. During the meeting, the code author makes sure feedback is elicited from the meeting members. Clarity of presentation and documentation of issues identified should occur. After the meeting, the author will review the identified issues and follow up with those attendees. This approach introduces corrections and other ideas generated from the walk-through.

Walk-throughs are often used for document reviews and include:

1. Functional requirements
2. Hardware requirements
3. Verification plans
4. Software requirements specification
5. Software design description
6. Software test documentation
7. Software user documentation
8. Test descriptions
9. Source code
10. Maintenance manual

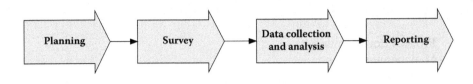

Figure F.1   One approach to an audit process.

11. System build procedures
12. Installation procedures
13. Release notes

F.1.5   Audits

The audit is an independent evaluation of the software performance and conformance to specifications, regulations, and standards. This review should be independent from the software development team to provide an *objective* review of the software. At the very minimum, the majority of the auditing staff are not members of the organization providing the software. This sort of review is equally applicable to hardware. Audits are sometimes performed by an external organization (in some cases, the auditor is the customer).

An example of the audit process (Figure F.1) is described below:

1. **Entry evaluation**—the initiator authorizes the start of the audit when the identified entry criteria have been met.
2. **Management preparation**—management ensures that the audit will be appropriately staffed with the required resources, time, materials, and tools, and that the audit will be conducted according to policies, standards, professional or legislative requirements, or other relevant criteria.
   a. Contractual obligations
   b. Organizational policies
   c. Standards
   d. Professional guidelines
   e. Legal requirements
3. **Planning the audit**—the initiator identifies the scope and confirms the objectives of the audit, identifies the individual auditors' areas of responsibilities and activities, and ensures the team is equipped.
4. **Kickoff meeting**—meeting conducted by the initiator to ensure that all parties have a common understanding and agreement on the audit objectives, responsibilities, and methods.

5. **Preparation**—auditors prepare all relevant materials for the audit examination review and take notes.
6. **Examination**—auditors conduct the audit by collecting evidence of conformance which is determined by interviewing the staff. These interviews include examining documents and witnessing processes, which are documented and analyzed. The lead auditor conducts a closing meeting with management, which compares the audit plan and measuring criteria to the audit results. The lead auditor prepares an audit report and delivers it to the initiator.
7. **Follow-up**—the audited organization (and the initiator) then act upon the result of the audit report.
8. **Exit evaluation**—the initiator provides indication that the audit process is complete.

Audits are often performed to verify the following:[3]

1. Backup and recovery plans
2. Contracts
3. Customer or user representative complaints
4. Disaster plans
5. Hardware performance plans
6. Installation plans
7. Installation procedures
8. Maintenance plans
9. Management review reports
10. Operations and user manuals
11. Acquisition and contracting methods
12. Reports and data (review, audit, project status, and anomaly reports)
13. Request for proposal
14. Risk management and contingency plans
15. Software configuration management plans
16. Software design description
17. Source code
18. Software project management, safety, and quality assurance plans
19. Software requirements specification
20. Software test documentation
21. Software user documentation
22. Software verification and validation plans
23. Technical review reports
24. Vendor documents
25. Walk-through reports
26. Durable media

Chapter Notes

[1]IEEE Standards Software Engineering, IEEE Standard for Software Reviews (Std 1028-1997), (New York, IEEE, 1998), p1.

[2]IEEE Standard for Software Reviews (Std 1028-1997), (New York, IEEE, 1998), p5.

[3]IEEE Standard for Software Reviews (Std 1028-1997), (New York, IEEE, 1998), p25.

Appendix G

Preliminary Design Review (PDR)

The preliminary design review (PDR) is conducted relatively early in the detailed development phase (see Figure B.1). The PDR is a technical review by the project-required disciplines (mechanical engineer, electronics engineer, etc.). This review gauges the ability of the proposed systems to meet the project needs (risk, cost, and delivery targets) and product needs (product cost, performance, and function targets). This activity includes review of physical and functional interfaces and possible configurations. The scope of the design review is not restricted to these design items. Additionally, the review critiques the proposed design against requirements and assesses the level of technical risk from production or manufacturing methods and processes.

In general, completing the PDR provides the project with:

1. Established system allocated baseline
2. Updated risk assessment for system development and demonstration
3. Updated cost analysis based on the system allocated baseline
4. Updated program schedule including system and software critical path drivers
5. Approved product support plan with updates applicable to this phase.

For complex systems, the program manager may conduct a PDR for each subsystem. The sum of the individual PDRs does not remove the need to perform an overall system PDR. When individual reviews have been completed, the overall system PDR will focus on functional and physical interface design, as well as overall system design requirements. The PDR determines whether the hardware, human, and software preliminary designs are complete, and whether the integrated product team is prepared to start detailed design and test procedure development.

For software portions of the project, this review will focus on:

1. Evaluating the development progress
2. Assessing consistency and technical adequacy of the selected top-level design
3. Evaluating the test approach
4. Assessing software requirements meets preliminary design

The PDR covers the following areas:

1. Hardware items
2. Software items
3. Support equipment
4. Evaluation of electrical, mechanical, and logical designs
5. Electromagnetic compatibility
6. Design reliability
7. Design maintainability
8. Human factors
9. System safety
10. Natural environment
11. Equipment and part standardization
12. Value engineering
13. Transportability
14. Test
15. Service parts and customer-furnished property
16. Packaging
17. Technical manuals

The program manager should conduct the PDR when all major design issues have been resolved. If the PDR confirms, then work can progress on the detailed design. The PDR should address any resolved critical, systemwide issues. Successful PDR means the questions below are able to be answered to the project team's satisfaction:

1. Does the state of the technical effort and design indicate the operational test and effectiveness success?
2. Does the present preliminary design satisfy the design requirements (is the design capable)?
3. Has a systems allocation baseline been defined and sufficiently documented to allow the detailed development to successfully proceed?
4. Are adequate processes and project and product metrics clearly defined and in place to support program success?
5. Do software requirements meet preliminary design?

6. Are ergonomic considerations appropriately addressed?
7. Is the program schedule realistic and executable (are technical risks too high)?
8. Are staffing levels adequate and appropriate?
9. Does the updated cost estimate fit within the identified budget?
10. Can the design be produced within the production budget and processes?
11. Is software functionality in the allocated baseline supported by software metrics?

Appendix H

Critical Design Review (CDR)

The critical design review (CDR) is conducted near the middle to the latter part of the detailed development phase of the project (see Figure B.1). The CDR is conducted to establish the suitability of the proposed unit design before coding and testing. In its most fundamental form, the CDR identifies software and hardware documentation that will be released for coding and testing activities (final development).

Like the preliminary design review (PDR), this review is a multidiscipline review of the system to gauge suitability for the project moving to the fabrication, demonstration, and test phase. Also like the PDR, this critique covers system performance, cost (project and product), as well as risk identification, including any unwanted system limitations. This review assesses the system's final design as defined in product specifications (product baseline). These product specifications enable the fabrication of the product and consist of textual specifications as well as drawings. Reviewed software designs must contain software design description (SDD).

The CDR occurs when the detailed design is complete. The purpose of the review is to

1. Determine if detailed design satisfies the performance and speciality engineering requirements
2. Assess the detailed design for compatibility among the hardware and software items such as:
   a. Associated interfacing equipment (subsystems and tools)
   b. Computer software
   c. Personnel

3. Assess software and hardware item risks
   a. Technical risks
   b. Project cost
   c. Product cost
   d. Schedule risk
4. Assess producibility of the design
5. Assess whether or how software requirements meet preliminary design
6. Determine the level of completeness of support and operational documents

The CDR covers the same areas as the PDR, but with the intent of closure in mind:

1. Hardware items
2. Software items
3. Support equipment
4. Evaluation of electrical, mechanical, and logical designs
5. Electromagnetic compatibility
6. Design reliability
7. Design maintainability
8. Human factors
9. System safety
10. Producibility and manufacturing
11. Natural environment
12. Equipment and part standardization
13. Value engineering (life cycle cost)
14. Transportability
15. Test
16. Service parts and customer-furnished property
17. Status of the quality assurance activities
18. Packaging
19. Technical manuals

Appendix I

Test Readiness Review (TRR)

The test readiness review (TRR) is a multidisciplined technical review to ensure that the subsystem or system under review is ready to proceed into the formal test (see Figure B.1). Testing a component, subsystem, or system usually consumes a significant amount of the project budget. The test readiness review objective is to ensure that the product test requirements (developmental as well as operational) will be fulfilled adequately. This review critiques software test procedures for completeness. Test procedures are evaluated for compliance to test plans and test descriptions, and adequacies for accomplishing the test objectives or verification of the product requirements. This review may also encompass reviews of informal software testing and documentation updates. A TRR is considered successful when the software test procedures and results are satisfactory, allowing the project to proceed to the formal test activity.

The TRR accomplishes the following:

1. Assesses test objectives
2. Covers any design or requirements changes
3. Confirms completeness of test methods and procedures and compliance with test plans and descriptions (identifies limitations)
4. Plots the scope of tests
5. Identifies safety issues pertaining to tests
6. Elicits known software bugs
7. Confirms that required test resources have been properly identified
8. Verifies test coordinating activities to support planned tests
9. Verifies traceability of planned tests to program requirements/ documented user needs

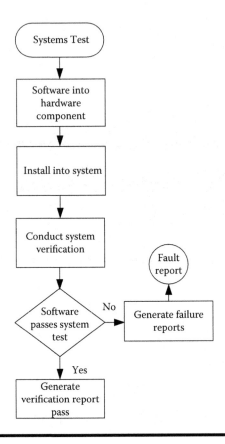

Figure I.1 System verification.

10. Assesses system under review for development maturity
11. Assesses risk to determine readiness to proceed to formal testing

In addition to adequate planning and management, to be effective the program manager must followup with the outcomes of the TRR. Items identified as risk areas must be resolved. While it is true that testing is not a quality-assuring activity, it is key to program success that problem areas are identified in advance of the customer receipt. The program manager should carefully plan and properly staff tests (Figure I.1). Test and evaluation is an integral part of the development engineering processes of verification and validation.

The type of risk and risk severity (impact) will vary as a system proceeds from component level to system level to systems of system level testing. Early component level test may not require the same level of review as the final system level tests. Sound judgment should dictate the scope of a specific test or series of tests.

1. Requirements changes
2. Design changes
3. Hardware/software test plans and descriptions
4. Hardware/software test procedures
5. Hardware/software test resources
6. Test limitations
7. Scheduling
8. Documentation updates

At the conclusion of the TRR, the following questions should be answered:

1. Why are we testing?
   a. What is the purpose of the planned test?
   b. Does the planned test verify a requirement directly traceable back to a system specification or program requirement?
2. What are we testing (subsystem, system, system of systems, other)?
   a. Is the configuration of the system under test sufficiently mature, defined, and representative (i.e., random size with enough material to make a judgment) to accomplish planned test objectives and support defined program objectives?
3. Are we ready to begin testing?
   a. Have all planned preliminary, informal, functional, unit level, subsystem, system, and qualification tests been conducted, and are the results satisfactory?
4. What is the expected result and how can (or will) the test results affect the program?
5. Is the planned test properly resourced (people, test article or articles, facilities, data systems, support equipment, logistics, etc.)?
6. What are the risks associated with the tests and how are they being mitigated?
7. What is the fallback plan should a technical issue or potential show-stopper arise during testing?

Upon completing the TRR, distribute review meeting notes. Typical success criteria for TRRs are

1. Completed and approved test plans for the devices, subsystem, and system under test
2. Complete identification of test resources and required coordinating activities
3. Assessment that previous component, subsystem, and system test results form a satisfactory basis for proceeding into planned tests
4. Identified risk level acceptable to the program leadership

Appendix J

Functional Configuration Audit (FCA)

Like the other audits, if the system is too complex, this may require an iterative audit. The functional configuration audit (FCA) is a formal audit to validate the development activity has been completed satisfactorily and that the configuration (software and hardware) under scrutiny has achieved the performance and the functional characteristics specified (see Figure B.1). The FCA is often a prerequisite to acceptance of the software and hardware.

The FCA is conducted on:

- Production proposed level of functionality
    - Preproduction
    - Prototype representative of production
- First production article

Review material is prepared before the FCA. The audit scope must be clear, which is accomplished by identifying the revision level of the specification and the software as well as a list of deviations or engineering change requests that pertain at the time of the review.

Test procedures and results (data) are to be reviewed for compliance with specification requirements. The following testing information should be available for the FCA:

- Test plans, specifications, descriptions, procedures, and reports for the configuration item;
- A complete list of successfully accomplished functional tests during which pre-acceptance data was recorded;

- A complete list of successful functional tests if detailed test data are not recorded;
- A complete list of functional tests required by the specification but not yet performed (to be performed as a system/subsystem test);
- Preproduction and production test results.

When software viability cannot be determined without system or integration testing, the FCA will not be considered complete until these activities have been concluded and assessed.

In addition to reviewing the test results, the FCA often reviews such items as any software user manuals, operator manuals, and any system diagnostics manuals. This review often uses initial production level parts, but can also use prototype parts that represent production.

After the FCA has concluded, the supplier provides a copy of the minutes from the review and the project manager on the customer side notes that the FCA obligation has concluded.

Appendix K

Physical Configuration Audit (PCA)

The physical configuration audit (PCA) is a formal examination that verifies the component and software as-built meets the technical documentation (see Figure B.1). A representative number of drawings and associated manufacturing instruction sheets for each item of hardware should be reviewed for accuracy to ensure that they include the authorized changes reflected in the engineering drawings and the hardware. The purpose of this review is to ensure the manufacturing instruction sheets accurately reflect all design details contained in the drawings. Since the hardware is built in accordance with the manufacturing instruction sheets, any discrepancies between the instruction sheets and the design details and changes in the drawings will also be reflected in the hardware. The following minimum information shall be recorded for each drawing reviewed:

1. Drawing number/title (include revision letter)
2. Date of drawing approval
3. List of manufacturing instruction sheets (numbers with change letter/titles and date of approval) associated with this drawing
4. Discrepancies/comments
5. Select a sample of part numbers reflected on the drawing. Check to ensure compatibility with the program parts selection list, and examine the hardware configuration item (HWCI) to ensure that the proper parts are actually installed

As a minimum, the following inspections should be completed for each drawing and associated manufacturing instruction sheets:

1. Drawing number identified in manufacturing instruction sheet should match latest released drawing
2. List of materials on manufacturing instruction sheets should match materials identified in the drawing
3. All special instructions identified in the drawing should be on the manufacturing instruction sheets
4. All dimensions, tolerances, finishes, etc., listed on the drawing should be identified in the manufacturing instruction sheets
5. All special processes called out in the drawing should be identified in the manufacturing instruction sheets
6. Nomenclature descriptions, part numbers, and serial number markings shown on the drawing should be identified in the manufacturing instruction sheets
7. Review drawings and associated manufacturing instruction sheets to ascertain that all approved changes have been incorporated into the configuration item
8. Check release record to ensure all drawings reviewed are identified
9. Record the number of any drawings containing more than five outstanding changes attached to the drawing
10. Check the drawings of a major assembly/black box of the hardware configuration item for continuity from top drawing down to piece-part drawing.

The supplier should establish that the configuration being produced accurately reflects released engineering data. This includes interim releases of service parts built before the PCA to ensure delivery of current service parts. The customer might audit the supplier's engineering release and change control system to ascertain that they are adequate to properly control the formal release of engineering changes. The supplier's formats, systems, and procedures should be used. The following information should be contained in the records:

1. Serial numbers, top drawing number, specification number;
2. Drawing number, title, code number, number of sheets, date of release, change letter, date of change letter release, engineering change order (ECO) number.

The contractor's release function and documentation will be capable of determining:

1. The composition of any part at any level;
2. The next higher assembly using the part number, except for assembly into standard parts;

3. The design of the part number with respect to other part numbers;
4. The associated serial number on which subordinate parts are used (if serialized);
5. Changes that have been partially or completely released against the item;
6. The standard specification number or standard part number used within any nonstandard part number;
7. The supplier specification document related to subsupplier part numbers.

The engineering release system and associated documentation shall be capable of:

1. Identifying changes and retaining records of superseded configurations formally accepted by the contracting agency;
2. Identifying all engineering changes released for production incorporation. These changes shall be completely released and incorporated before formal acceptance of the configuration item;
3. Determining the configuration released for each configuration item at the time of formal acceptance.

Engineering data should be processed through a documentation control department to ensure coordinated action and preclude individual release of data when part number coordination is required. Engineering change control identifiers should be unique. Parts that have demonstrated compliance with the product specification are approved for acceptance as follows: the PCA team certifies that the part has been built in accordance with the drawings and specifications.

After completion of the PCA, any future changes will be handled via engineering change requests.

Appendix L

Formal Qualification Review (FQR)

The formal qualification review (FQR) includes the test, inspection, or analytical techniques and processes employed to assess a set of software and hardware items that comprise the system. These items meet specific performance and functional requirements. This review can take place simultaneously with the functional configuration audit (FCA); however, if qualification is deemed unobtainable without completion of system testing, the FQR can wait until completion of systems testing.

Qualification is not limited to the following, but it can include:

1. Waiver/deviation list prepared
2. Qualification test procedures submitted
3. Qualification testing completed
4. Qualification test results compiled and available
5. Facilities for conducting FCA available
6. Qualification test procedures reviewed and approved
7. Qualification testing witnessed
8. Qualification test data and results reviewed and approved

Upon completion of the FQR, the supplier distributes the meeting minutes to the participants and to the customer.

Appendix M

Production Readiness Review (PRR)

The production readiness review (PRRs) objective is to determine if production capabilities are sufficient to be able to produce the product. This is done through reviews of the actions identified earlier in the program to deliver the production processes for the product. This is often an iterative review, with the level of refinement (details) of the reviews increasing as the reviews progress. Typically, early reviews will tackle low-yield manufacturing processes as well as manufacturing development issues to produce a design that meets requirements. The latter reviews can encompass production planning issues as well as facilities concerns.

Typical PRR success criteria (*Defense Acquisition Handbook*) include affirmative answers to the following exit questions:

1. Has the system product baseline been established and documented to enable hardware fabrication and software coding to proceed with proper configuration management?
2. Are adequate processes and metrics in place for the program to succeed?
3. Are the risks known and manageable?
4. Is the program schedule executable (technical/cost risks)?
5. Is the program properly staffed?
6. Is the detailed design producible within the production budget?

A follow-on, tailored, PRR may be appropriate in the production and deployment phase for the prime contractor and major subcontractors if:

1. Changes occur from the system development and demonstration phase and during the production stage of the design, in either materials or manufacturing processes;
2. Production startup or restart occurs after a significant shutdown period;
3. Production startup with a new contractor;
4. Relocation of a manufacturing site.

Appendix N

Embedded Development Overview

Embedded software development requires extensive planning, monitoring, and thorough follow-up to succeed in all quarters. The responsibility of the project manager is to prepare the schedule for documentation, development, build, and quality assurance. In order to manage all of these activities and risk, an understanding of the workflow is necessary. The workflow is essentially a model of the production sequence for designing, implementing, and testing embedded software—a map of the process.

General embedded development knowledge and the workflow model allow the project manager to ask pertinent questions. These questions provoke answers about risk from those responsible for the deliverables. Moreover, it allows the project manager to do some risk assessment on his or her own. Knowledge of acceptable software life cycle processes (via the workflow model) makes it possible to identify situations of increased risk.

There are a number of development cycle models. The project team should reference the same model since this provides a common frame of reference for subsequent team discussions.

1. Waterfall model
2. Spiral model
3. Big bang model
4. Iterative
5. Evolutionary development
6. V-cycle model
7. Rapid prototyping
8. Model-driven

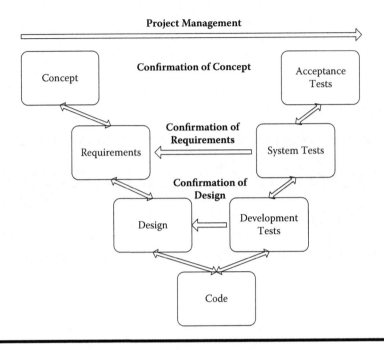

Figure N.1 Software "V" life cycle.

Figure N.1 is an illustration of the "V-cycle" model for software development life cycle. The model illustrates how the development starts from a systems level and progresses through increasingly detailed development phases. The activities start at the upper left-hand corner with customer requirements. This customer input drives concept selection and documentation. Once the team defines the concept, it generates the system requirements that support the concept. The system requirements are broken down to the next level of detail where the component's requirements are documented (hardware and software specifications). Finally, with the component level documentation complete, the hardware and software design implementation work starts. The output of each block is the input to the next one.

N.1 V-cycle Model

This model provides a juxtaposition of the development work (left side of the "V"), with the verification or testing work (right side of the "V") necessary to quality secure the product. In Figure N.1 each step has related verification to ensure the requirements are satisfied. For example, design is verified by developmental tests. The model applies to both software and hardware. During the hardware design stage, performance and functional

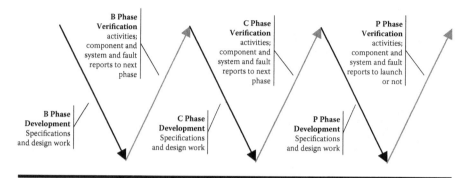

Figure N.2   Software "V" iterative life cycle.

theories and hypotheses can be tested (verified) to either substantiate or refute an expected quality level.

As with all models, there are limitations with this model. The "V-cycle" model does not adequately reflect the iterative nature of the development process. However, with some imagination it is possible to extend the V-cycle graphic to account for the additional phases (refer to Figure N.2).

N.2   Embedded Software Development Tools

A unique set of tools are required for developing embedded software.

N.2.1   Hardware Tools

Hardware tools are used to physically develop the component. These tools simulate or emulate the embedded system or measure the state of the hardware for debugging of the developed system.

- In-circuit emulator
- Simulators
- Logic analyzer
- Logic probes
- Voltmeters
- Oscilloscopes
- Schematic tools
- Printed circuit board (PCB) generators
- Waveform generators
- Data bus emulators
  - J1850
  - J1939

- LIN
- USB
■ System emulation

N.2.2 Soft Tools

Soft tools facilitate development work by focusing directly on software creation and documentation handling.

- Real time operating system (RTOS)
- Text editors (write software)
- Compilers
  - Assembly compiler
  - C compiler (C is the most common embedded software language)
- Software debugger
- Software files and data management tools
- Change management tools
- Fault or error management tools

N.3 Embedded Software Development Process

The software development life cycle involves the following stages of product development.

N.3.1 Functional Specification

The functional specification captures all the requirements in a predefined format so that the development team can deliver a product. Sometimes, the customer will deliver the functional specification. Once the customer requiremens are documented, the design team will generate design concepts in response to the specification. The concept with the highest probability of success—meeting both customer and the supplier needs—is selected. The concept selection can take the form of a gate event. Any development issues and clarification on the customer requirements should be discussed until a clear understanding evolves and the appropriate documents are updated in order to prepare the function test requirements.

N.3.2 System Development

Once the concept is selected, the team may create system-level specifications if that level of specification is appropriate to the project. When needed, this work may be aided by prototype parts to prove concepts and test interactions. The development team may create system-level state diagrams during the system development work. The state diagrams become descriptive models for the development of software modules. Completion of the state diagram makes it possible to apportion the functional demands through the proposed system.

N.3.3   Detailed Design

All the software requirements should be gathered in the software requirements specification (SRS) and should be reviewed by the customer before final development. This level of detail frequently uses data flow diagrams and mapping of the software functions to the system architecture. In most cases, complex software is logically grouped into different modules to enhance control of the software interfaces.

N.3.4   Final Development

N.3.4.1   Coding

Figure N.3 shows a typical software development process from specification to unit testing. Once a requirements review occurs, the engineers begin coding. Based on the module design, code occurs in the chosen embedded development language. This process could involve an RTOS and associated debuggers, compilers, and test tools to verify the output of these modules. Usually coding is associated with a unit test plan where individual modules are verified for their capability and defect-level before moving on to the next phase. This *unit test* may be conducted by the individual writing the software module.

N.3.4.2   Software Integration

During software integration, all software modules meld into a single package (see Figure N.4). Integration is the terminus of software development with the exception of defect removal and software maintenance.

During the coding phase, the development team writes test plans and procedures to exercise all of the features within the software build. With the build complete, the test team uses these documents to drive verification of the software. It executes component verification modules, schedulers, and the RTOS. System integration verifies the interaction among software modules and the interfaces between the software and hardware.

Upon passing the verification activities, the development team will publish the release notes. These release notes describe the revision level and feature set of the software build. The software is then either sent to customers to upload into the components on their site, or sent to the supplier

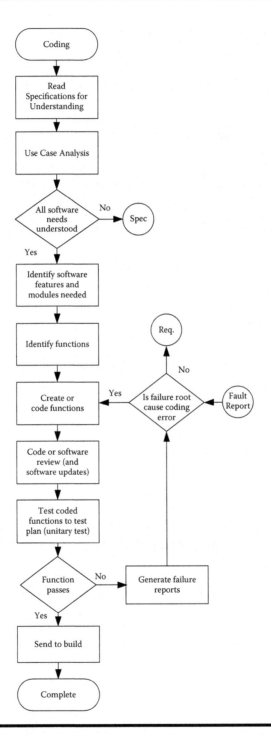

Figure N.3 Software coding process.

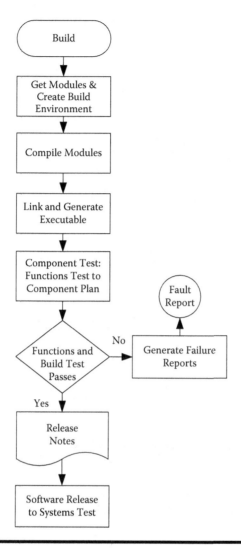

Figure N.4 Software (SW) build process.

responsible for integration into the hardware prior to sending to the customer.

If the software does not pass verification, discussions with the system tester ensue. If the performance and functional deviations are acceptable to the customer, the release notes are updated with known performance issues and the process is complete. If deviations are prohibited (discovered fault is too risky), then the development team will rework the software, which then delays the start of the systems verification activities.

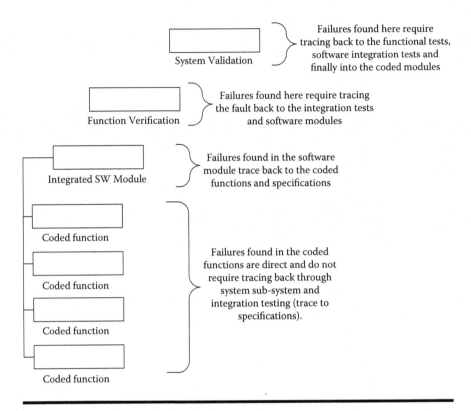

Figure N.5   Software debug hierarchy.

N.3.5   System Verification

System verification can be grouped into black box testing and white box testing. Black box testing occurs when the internal behavior of the module, software or hardware, is unknown. The test team takes the sample part or prototype and stimulates the inputs and records the outputs. If the test document is complete, the testers should be able to log anomalies for subsequent review with the development team.

White box testing occurs when the internals of either the software or the hardware are sufficiently well understood to drive detailed testing of internal features. These tests will determine the code coverage and the branch coverage of the particular system module (if using white box testing for software). A combination of black and white box testing provides for more complete software testing. The farther along in the project an error in software is detected, the more costly and time consuming the correction (see Figure N.5). The sooner testing starts and finds problems, the quicker solutions can be generated.

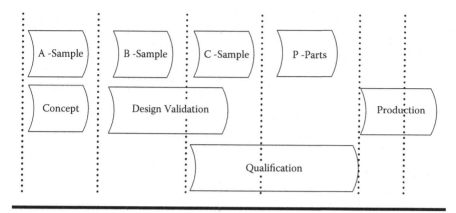

Figure N.6 Parts and phases.

N.4 Part Types

The following discussion provides on example of the use of sample part levels. While there are myriad ways to treat these parts, no one solution emerges. The part types change from organization to organization. During development, mock-ups, concept pieces, and limited functional parts are created for analysis by both supplier and customer. These parts make it easier to confirm that development is proceeding as planned and allows for customer feedback. The incremental approach reduces risk because the team solicits customer response at every release. Figure N.6 shows one way this system of parts delivery could work.

N.4.1 "A" Sample

The "A" designation applies to the initial prototype model. *"A" samples* provide part of the answer for a request for quote sent to multiple suppliers. The samples allow the customer to review the concepts proposed by suppliers for determination of the best possible candidate solution. These parts are usually stereolithographic models or similar type of model and cannot take much physical abuse. In its minimal form, this part type has no software installed and is useful for fit and appearance only. In a more sophisticated form, the "A" sample has enough features to provide the basis for the later development and implementation of the product.

"A" samples are not used with projects that

1. Reuse existing software,
2. Add features into an existing platform,
3. Use a large proportion of software from a previous project.

The time for developing the "A" sample should be used to create the following activities:

1. Determine product architecture (hardware, software, mechanics, and fit),
2. Prove feasibility,
3. Create early software system design.

"A" sample—risk The team can reduce the risk by

1. Checking for clarity of the specification,
2. Verifying the early system design,
3. Following a definitive plan software verification.

N.4.2 "B" Sample

The *"B" sample* is a workable version of the software and hardware. The concept is available in a refined form and is ready for initial prototype testing. From a hardware perspective, these parts usually represent production parts. However, although the "B" sample parts often use production material they will frequently build by a prototype operation rather than the final production line. The physical parts may be from a soft-tooled part, a short-lasting equivalent to the much more durable hard tool. Because the soft tool is cheaper, the design team can use such a tool to build prototypes to assess the feasibility of going to the hard tool.

If the development team has ventured into a new technology, it may create several iterations of the "B" samples. With each revision more features and corrections appear. The software versions contain an increasing functional capability and are optimally developed by passing through a predefined sequence of development and stabilization cycles. The objective is to implement the complete feature set of the product by the conclusion of this "B" sample period. The emphasis of activities is with implementing functional capability and testing.

The following conditions apply:

1. Software system design has been frozen and reviewed,
2. File and program structures are known and their sizes are calculable,
3. Execution and response times correspond to those on the target system.

"B" sample—risk Some risks to consider are

1. How well does the initial concept support the required feature set,
2. How are the verification and validation tests proceeding,

3. How gracefully is the team working with the customer to improve and clarify all functional requirements.

N.4.3 "C" Sample

This "C" sample software ultimately becomes the completed version if testing goes well. Like the "B" sample parts, the "C" samples are iterative. For these parts, however, the totality of the required feature set is complete. These parts are usually used for remaining systems integration tasks. The system verification activity provides evaluative information to the supplier for software correction and subsequent updating of the code. On completion of systems integration work and closure of any corrections, the software is ready for both the production and maintenance phase. The "C" sample parts use centers around defect removal and correction and support of preparation for production. Smaller modifications can be implemented during a short development cycle, but development consists mainly of product stabilization. The software of the "C" sample is used for acceptance testing. Since these parts are so close to production intent, many organizations use these parts to get some field and customer test and evaluation.

The following conditions apply during the "C" sample activities:

1. All functions in the software requirements specification are fully realized
2. Software testing has concluded successfully.

Production preparation   During product preparation the production and assembly lines are prepared and the production tools are fabricated.

Qualification   Qualification of the product samples manufactured under series conditions. Qualification includes software qualification testing. The customer has to decide about series release. Qualification is the responsibility of the department for quality and test.

"C" parts—risk   The team controls risk in the "C" parts by

1. Checking that all functional requirements are covered,
2. Executing a quality check,
3. Verifying progress status,
4. Following up with meetings to correct issues.

N.4.4 "P" Sample

The "P" sample parts represent the teminus of development activities and the commencement of the maintenance phase for the software. The "P" sample parts are built under production conditions and delivered per contract. All verification activities are over and the product is ready for the marketplace.

N.5 Risks in Software Development

Test equipment is completely available   All the test equipment needed to verify the software should be installed and validated. Appropriate tools like emulators and debuggers should be available no later than the planning phase to ensure that the schedules and budget remain on track.

Hardware and software developers do not communicate and are not compatible   There are many instances during which the hardware and software developers proceed on the assumption that both sides understand the requirements. Any changes must be communicated to both hardware and software developers to ensure that no piece of the product will have problems in the future due to the change of one of its features— sometimes ironically called "collateral damage." This problem becomes even more pronounced and the risk higher when dealing with multiple software configuration items from differing organizations; for example, more than one supplier may be providing the embedded software for the project.

Major change requests occur during the final stages of release   Changes should be minimized during the final stages of development. Any changes from the initial concept should be merged as early as possible in the development cycle. Late inception of major changes brings complication to the software, risking the quality of the deliverable. Frequently, project software engineers are starting to be reassigned to other projects. This resource reduction makes the code modification especially difficult to plan and accomplish.

Well-defined requirement specification   Many of the problems with embedded software development arise because of incomplete, improperly stated, or misunderstood requirements. It is very important to have all the requirements clarified in the initial stages of project discussion. Additionally, requirements reviews between the supplier and the customer technical staff increase the probability of high-quality requirements. The more organizations are involved in the development effort, the more critical the need for reviews.

No project schedule baseline kept   Project management requires planning and comparing actual performance to the plan to enable control of the project. The project manager and project team create project schedules but do know how to ensure that project schedule baselines are kept. It is very difficult to predict and anticipate the project schedule without adequate commitment even at the detail level.

Undefined change management process   Uncontrolled change is a quick way to reduce the probability of project success not only during the late stages of a project. Uncontrolled change often means it is undocumented and with no concordance among stakeholders. If there is no procedure for making changes to the product, any person can change it at any time. The result is documents that are unsynchronized with the development process. The ripple effect of uncontrolled change affects verification plans and test descriptions, which will not be updated appropriately.

Automation of tests   Automation of the test process can improve the timeliness of testing. Complex and long tests are good candidates for automation. Automated testing can save time, money, and human resources needed for the project. Time spent clarifying the automated testing process offers a chance for improving the product requirements. If the embedded software development team uses a test, analyze, and fix (TAAF) approach to development, quick turnaround using automated testing allows for significantly more iterations of the TAAF sequence.

Automated testing and the software required to accommodate such testing create another software synchronization requirement. Any product software change in the product will generate a need for a change within the automated test system. To make best use of automated testing, the test software has to be *available and qualified before the software under test becomes available,* in effect making our approach analogous to test-driven development. If the automated test software has not been verified, it is not possible to make assessments about the component software under test.

Appendix O

Process Sign-Off

Different organizations may require different information contained within the process sign-off (PSO) document. In many cases, PSO is a formal document that might include the following items:

1. Part number and change level
   a. Bills of materials
   b. Drawings
   c. Engineering change notifications
2. Process flow diagram and manufacturing floor plan
   a. Process quality control
   b. Rework handling instructions
   c. Manufacturing floor plan
3. Design FMEA and process FMEA
   a. Fault tree analysis
   b. DFMEA
   c. PFMEA
4. Control plan
5. Incoming and outgoing material qualification and certification plan
   a. Goods receiving instructions
   b. Receiving inspection instructions
   c. Receiving inspection tracking for components
6. Evidence of product specifications
   a. Drawings and dimensional information
   b. Dimensional checklist

7. Tooling, equipment, and gauges identified
   a. Equipment list
      i. Equipment name
      ii. Equipment number
   b. Equipment calibration due
8. Significant product and process characteristics identified
9. Process monitoring and operation instructions
   a. Instructions for each process
      i. Process name
      ii. Part number
      iii. Issue date
      iv. Revision
   b. Tell tales for process breakdown for each process
10. Test sample size and frequency
    a. Lot acceptance sample table
    b. Product qualifications requirements
       i. Sample results
       ii. Actions on process
       iii. Actions on lot
11. Parts handling plan
    a. Antistatic handling procedures
    b. Material handling on the line
    c. Tracking materials on the line
    d. Stores control
       i. Objectives
       ii. Scope
       iii. Responsibilities
       iv. Procedure
       v. Date, revision level
       vi. Sign-off
    e. Nonconforming material handling
       i. Objectives
       ii. Scope
       iii. Responsibilities
       iv. Procedure
       v. Date and revision level
       vi. Sign-off
    f. Nonconforming material tag (attached to failed component)
       i. Tracking tag number
       ii. Part name
       iii. Carton date
       iv. Packer number
       v. Reason and corrective actions required
       vi. Corrective action

       vii. Date/time

      viii. Quality and production sign-off

   g. Nonconforming material report

        i. Tag number

       ii. To/from

      iii. Project or program name

      iv. Description of discrepancy

       v. Quantity received

      vi. Quantity rejected

      vii. Disposition

         A. Returned for evaluation

         B. Use as is

         C. Other (specify)

         D. Returned for rework

         E. Reworked at customer or supplier expense

      viii. Approvals

         A. Materials approval

         B. Manufacturing approval

         C. Quality assurance approval

       ix. Supplier feedback

         A. Cause of discrepancy

         B. Corrective action

         C. Date discrepancy will be corrected

         D. Preventative actions

         E. Containment

         F. Supplier sign-off

12. Parts packaging and shipping specifications

   a. Material handling requirements

        i. Part number

       ii. Part description

      iii. Supplier name

      iv. Manufacturing location

   b. Container description

        i. Container dimensions

       ii. Packaging weight

      iii. Quantity per container

      iv. Packaging cost per part

       v. Packaging usage

      vi. Interior dunnage

13. Product assurance plan (PAP)

14. Engineering standards identified

   a. General product description

   b. Functionality

   c. Key specifications

15. Preventative maintenance plans
    a. Goals of preventative maintenance
    b. Preventative maintenance chart
        i. Equipment number
        ii. Equipment name
        iii. Maintenance period (weekly, every 6 months, etc.)
        iv. Date
        v. Comments
    c. Preventative chart for special equipment
16. Gauge and test equipment evaluation
    a. Gauge equipment identified
    b. Gauge reproducibility and repeatability (GR&R) results
    c. Equipment capability studies
17. Problem-solving methods
    a. List methods for solving problems
18. Production validation complete
    a. Design verification plan and report (DVP&R)
19. Buzz, squeak, rattle (BSR) and/or noise, vibration, harshness (NVH)
20. Line speed demonstration and capability evaluation
21. Error and mistake proofing

Much of this material accumulates during development; however, the final document relies on results from production start. Often, this document is reviewed during the run-at-rate phase. The output of this review is added to the production capability information (Cpk) generated from the production line.

Appendix P

Software Quality Metrics

Well-chosen software metrics such as *defect density* and *failure rates* provide objective information about the condition of the software. Good metrics provide the project manager with information useful for determining the feasibility of embedded software release.

P.1 Overview

Software metrics are performance based and are sometimes specified by the customer. Some examples of useful metrics are as follows:

- Lines of code (if using the same computer language for all embedded software)
- Function points
- Defects per line of code
- Absolute defect count
- Defect per time quantum (for example, defects/week)
- Halstead metrics
- McCabe's cyclomatic complexity calculation

P.2 Need Driven

Projects can have numerous stakeholders, each with demands and opinions on what constitutes a successful product. For a metric to be useful, there must be a customer for the metric and the metric must be capable of calculation by an experienced developer; that is, sufficiently objective that any trained developer would produce the same result.

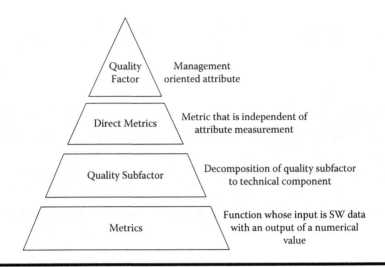

Figure P.1   Software metric hierarchy.

P.3   Methodology

A detailed account of software quality metrics methodology can be found in IEEE standard 1061 "Software Quality Metrics Methodology." However, a summary of the approach appears below:

1. Establish software quality requirements
2. Identify software quality metrics
3. Implement software quality metrics
4. Analyze metric results
5. Validate software quality metrics

P.3.1   Software Quality Requirements

This phase identifies quality requirements from the respective of the project participants or stakeholders. This is not restricted to customer inputs, however; even customer inputs can derive from a large audience. Any participant who can have a valid claim to software quality expertise should participate or review the quality requirements. Quality attributes that conflict with other attributes or requirements must be resolved. The software quality attributes are then prioritized and quantified. As it is with all requirements, they must be clear, verifiable, and measurable.

Examples of software quality requirements could be

■ The processor load must be below 80 percent.
■ The memory use must be less than 75 percent of total memory (future functionality)

- The system response to a key press input should be less than 100 milliseconds
- Product must not suffer from a "locked up" condition

The U.S. Department of Defense provides a list of measurement criteria in the 1994 publication of *Software Development and Documentation*, MIL-STD-498. This publication outlines measurement areas as:

- Requirements volatility
- Software size
- Software staffing
- Software complexity
- Software progress
- Problem change report status
- Build/release content
- Computer hardware resource utilization
- Milestone performance
- Scrap and rework
- Effect of reuse

P.3.2   Software Quality Metrics

The goal of this phase is to produce an approved list of quality metrics, which can be done by using the quality requirements list as input. The team will perform a cost benefit analysis on each of these metrics to determine if the cost for using the metric produces a corresponding or greater benefit.

The following are some simple software quality metrics:

- Software defect density (number of known defects/code size),
- Problem arrival rate,
- Problem resolution rate (how quickly are we fixing the problems?),
- First-pass yield.

P.3.3   Implementation

During this phase, the team quantifies the data to be gathered supported by the metrics from the prior phase. This produces the data gathering restraints, methods of collection, traceability requirements, as well as any training and associated schedule.

The metrics are only as good as the collection method and source data that are used to create them. The process for gathering (tools, techniques, and evaluation procedure) requires handling of the agreed-on root data to ensure that it is not corrupted or compromised.

Data description, per IEEE standard:

Table P.1   Table of Software Metric Attributes

Item	Description
Name	Name given to the data item
Metrics	Metrics that are associated with the data item
Definition	Straightforward description of the data item
Source	Location of where the data item originates
Collector	Entity or individual responsible for collecting the data
Timing	Time(s) in the life cycle at which the data item is to be collected
Procedures	Methodology (e.g., automated or manual) used to collect the data
Storage	Location of where the data are stored
Representation	Manner in which the data are represented, e.g., precision and format
Sample	Method used to select the data to be collected and the percentage of the available data that is to be collected
Verification	Manner in which the collected data are to be checked for errors
Alternatives	Methods that may be used to collect the data other than the preferred method
Integrity	Person(s) or organization(s) authorized to alter the data item and under what conditions

P.3.4   Analyze Metric Results

This activity compares the data from the previous phase with the targeted data identified in the implementation phase. Disparities resulting from this comparison are investigated and analyzed and root cause and corrective actions occur. This corrective action could include modifying the metric.

P.3.5   Validate Software Quality Metrics

This validation is performed using statistical analyses on the data. This is a mathematically intensive process and not really the subject for a project manager, but rather intended for the enterprise statistician or, possibly, a six sigma black belt. The important concept is to understand the general methodology and benefits of software metrics. The results are reviewed by the project team, the customer, and assorted stakeholders.

Appendix Q

Project Metrics

Project metrics are different from software quality metrics. These measures are an attempt to determine the project status. Project metrics should allow for more specific performance failings to be identified and corrective actions used to address inadequacies. Common choices are the following items:

1. Budgeted cost work performed
2. Actual cost work performed
3. Budgeted cost work scheduled
4. Cost performance index (CPI)
5. Schedule performance index (SPI)
6. Estimate at completion (EAC)
7. Budget at completion (BAC)

The measurements are identified by the project as necessary to the success of the project phase. We include some nonexhaustive lists of project metrics by project phase.

Q.1   Voice of Customer

Metrics for this activity are centered around the understanding and capturing of the product and project requirements. This activity includes details such as specification writing and reviews.

1. Number of specifications needed
2. Number of specifications completed
3. Number of specifications reviewed
4. Ratio of specifications reviewed to total specifications needed
5. Number of requirements
6. Number of requirements reviewed

7. Number of new requirements or changes
8. Ratio of open action items to closed action items

Q.2   Product Development

Metrics for this phase are centered around the details of the product development. This includes details around function creation and hardware development. Consult the relevant chapter for details.

1. Number of software (hardware) functions needed
2. Number of software (hardware) functions completed
3. Number of software (hardware) functions reviewed
4. Ratio of software (hardware) functions reviewed to total software functions needed
5. Number of development tests planned
6. Number of development tests conducted
7. Number of failures or faults reported from development tests
8. Status of development test failures reported
   a. Under investigation
   b. Resolved
   c. Under verification
   d. Closed
9. The severity of development test failures reported
   a. Minor effect
   b. Appearance
   c. Mission critical
10. Average time of resolution of faults or failures from development tests
11. Actual staff levels compared to targeted staff levels
12. Ratio of open action items to closed action items

Q.3   Process Development

Metrics for the process development phase are centered around the deliverables from the production processes and production line requirements. This includes measurements of activities that are found in the relevant chapter.

1. Number of process instructions needed
2. Number of process instructions reviewed
3. Number of process instructions completed
4. Ratio of process instructions reviewed to total process instructions needed
5. Process flow completeness
6. Process flow plan completeness

7. Ratio of open action items to closed action items
8. Actual staff levels compared to targeted staff levels
9. First-pass yield to expected yield

Q.4   Verification

The metrics for verification focused on the deliverables for this phase tie in closely with the failure reporting, analysis, and corrective action system (FRACAS) system. The verification team uses this activity to detect errors and faults and report this information to the development team. Understanding the problem arrival rate and the subsequent corrective actions allows the project manager and team to make an assessment as to whether the actions to improve the quality of the project are successful. The whole process allows for some prediction regarding the risks to the project.

As an example, if we have an input fault rate trend of ten per week and a settlement of five per week, it suggests little likelihood of being able to meet the project and product quality delivery in the immediate future. In short, we would never catch up!

1. Number of test cases needed for the test plan (planning phase)
2. Number of test cases constructed at present (planning phase)
3. Number of faults reported
4. Number of tests conducted
5. Total number of tests to be conducted
6. Severity distribution of faults
    a. Mission critical
    b. Quality perception
    c. Cosmetic
7. Resolution distribution of faults
    a. Investigation
    b. Solution pending
    c. In test
    d. Resolved in release
8. Average time to close faults
9. Ratio of open action items to closed action items
10. Average number of faults reported per unit time
11. Actual staff levels compared to targeted staff levels

Appendix R

IEEE-1220 Systems Engineering Master Plan Format

The general layout of the systems engineering master plan (SEMP) is as follows:

1. Scope
2. Applicable documents
3. Systems engineering process application
   a. Systems engineering process planning
      i. Major deliverables and results
         A. Integrated database
         B. Specifications and baselines
      ii. Process inputs
      iii. Technical objectives
      iv. System breakdown structure
      v. Training
      vi. Standards and procedures
      vii. Resource allocation
      viii. Constraints
      ix. Work authorization
   b. Requirements analysis
   c. Requirements baseline validation
   d. Functional analysis and allocation
   e. Functional architecture verification
   f. Synthesis

    g. Physical architecture verification

    h. Systems analysis

        i. Trade studies

        ii. System/cost effectiveness analyses

        iii. Risk management

    i. Control

        i. Design capture

        ii. Interface management

        iii. Data management

        iv. Systems engineering master schedule (SEMS)

        v. Technical performance measurement

        vi. Technical reviews

        vii. Supplier control

        viii. Requirements traceability

4. Transitioning critical technologies
5. Integration of the systems engineering efforts

    a. Organizational structure

    b. Required systems engineering implementation tasks

6. Additional systems engineering activities

    a. Long-lead items

    b. Engineering tools

    c. Design to cost

    d. Value engineering

    e. Systems integration plan

    f. Interface with other lifecycle support functions

    g. Other plans and controls

    h. Configuration management of SEMP

7. Notes

    a. General background information

    b. Acronyms and abbreviations

    c. Glossary

Appendix S

Release Notes

Release notes are provided by the software and embedded hardware suppliers. Release notes are the development organization's way of describing the status of the latest deliverable to the customer. This document typically contains:

1. The latest software or hardware revision number,
2. A running revision number history,
3. Functions contained in this revision of software (engineering change request or ECR numbers, function names, and function revision levels),
4. Repaired faults and failures reported from previous releases,
5. Latest faults and failures in this revision of software or hardware,
6. Restrictions for the software or hardware,
7. Functional tests performed,
8. Results of functional tests,
9. Special tools needed to program.

Release notes are an attempt to minimize the customer's negative surprise factor by clearly articulating what is contained within the software and/or hardware delivery. In complex systems, all functionality may not be delivered at once. This information allows the customer's development organization to optimize the way it works with the latest deliverable from the customer by defining any known problems and tracking the release revisions.

Appendix T

FMEA Basics

T.1 What Is an FMEA?

T.1.1 Formal Definition

The acronym "FMEA" corresponds to the words "failure mode and effects analysis." An alternate form is "FMECA" or "failure mode, effects, and criticality analysis," an extended form of the tool.

The Society of Qutomotive Engineers' SAE J1739 [SAE 1994] defines the FMEA to be " a systemized group of activities intended to: (a) recognize and evaluate the potential failure of a product/process and its effects, (b) identify actions which could eliminate or reduce the change of the potential failure occurring, and (c) document the process.

MIL-STD-1629A [MIL1980] says that an FMEA (FMECA) can serve to[1]

> systematically evaluate and document, by item failure mode analysis, the potential effect of each functional or hardware failure on mission success, personnel and system safety, system performance, maintainability, and maintenance requirements.

T.1.2 Standards

SAE J1739   SAE J1739 is the official SAE version of the automotive FMEA. The less official—but more commonly followed—version is the third edition of the Automotive Industry Action Group (AIAG) *Potential Failure Modes and Effects Analysis* book. The SAE standard typically lags behind innovations contained in the AIAG book.

MIL-STD-1629   The military standard for failure mode analysis is MIL-STD-1629A. It adds the additional concept of criticality analysis (e.g., the wing may not fall off often, but when it does, it is critical!). The format for a 1629 FMECA is also somewhat different than that for an automotive FMEA.

IEC-812   This standard is little used in the automotive industry, but it may ultimately represent a move by International Organization for Standardization/International Electrotechnical Commission (ISO/IEC) to define the FMEA internationally.

T.1.3   But, What Is It Really?

The standards use nice words to define the FMEA. In reality, however, what are we trying to do? In its essence, the FMEA is a simple matrix that captures potential problems and provides a convenient format for documenting solutions.

A practitioner can easily get lost in the "ins and outs" of the numerical portions of the FMEA. I have personally seen arguments about severity, occurrence, and detection rage for hours before team members grudgingly come to a compromise. Successful FMEA users will keep foremost in their mind the purpose of the FMEA: to "anticipate" problems so that designers and manufacturers can eliminate them before the product ever hits the market.

Other variants of the FMEA format—service FMEAs and help desks—are primarily documentary tools instead of anticipatory money savers. In essence, these documents use a convenient format that may or may not derive from the design FMEAs and the process FMEAs to assist the user in troubleshooting.

T.1.4   Similar Tools

MORT analysis   Management oversight and risk tree (MORT) analysis is a tool used by the Department of Energy for both anticipation and analysis of accidents. A MORT diagram uses "OR" and "AND" symbols to show the relationships among and between subordinate concepts. A postaccident MORT analysis may consist of a battery of 1,500 questions during the investigation. An alternative to the full MORT is the so-called mini-MORT, which is a condensed version of the regular tool.

Fault trees   Fault trees nearly always use some variant on the logical circuit symbols to represent "OR" and "AND" conditions. If we picture a three-year-old child constantly asking "why," we can come close to the method behind the fault tree.

Fault trees become complex very quickly, so we normally only see them in either very simple situations or in mission critical (safety-oriented) applications.

Fault trees backtrack from an incident or an imagined incident to a collection of potential causes. Difficulties arise when practitioners tend to consider prejudicially only "OR" situations or primarily "AND" situations. The "OR" is normally an exclusive "OR"—no "AND" condition applies. The "AND" condition requires that both causes be present before the next situation on the tree can occur.

T.1.5 The Philosophy

The overriding concern in the use of tools such as FMEA, quality function deployment (QFD), and fault trees is the ability to *anticipate* problems. A problem that we can anticipate is a problem that we can manage. The idea is that we use the tools, not that the tools use us.

The worst thing that can happen to any of these tools is that they become a mechanical, rigid march through the technique. "Going through the motions" suggests thoughtlessness, which is the antithesis of what these tools attempt to accomplish.

T.2   Why Do an FMEA?

T.2.1   Anticipation

One of the most significant benefits of the FMEA lies in the anticipation of failure. This does not mean we are compiling self-fulfilling prophecies—it means we are taking deliberate and systematic efforts to manage problems and risks before they can become problems and risks. And we capture our efforts in a compact format useful for quick study and conducive to terse descriptions.

T.2.2   Problems

Ongoing problems can be captured in an FMEA as "lessons learned." The FMEA is one format that can be used to systematically capture incidents and save them for consideration on the next round of product or process design.

T.2.3   Documentation

The FMEA can also serve as documentation of "due diligence." That is, should we have to go to court during a litigation, a well-formulated FMEA can serve as evidence that we have worked diligently to design or manufacture a high-quality product.

T.2.4 ISO/TS 16949

ISO/TS 16949 is the automotive version of the ISO 9001:2000 standard for quality systems. ISO/TS 16949 spells out the requirements for an automotive quality system, which is largely a superset of the ISO 9001 standard. FMEAs are a requirement under 16949.

T.2.5 Product Improvement

The design FMEA can be used to eliminate potential problems with the product ahead of development, during development, or after development. The best event, of course, occurs when the DFMEA is created early on.

T.2.6 Process Improvement

The process FMEA can be used to eliminate potential problems with the manufacturing process ahead of development.

T.3 When to Do an FMEA?

T.3.1 Early in Development

The earlier the design/process teams can construct an FMEA, the earlier they can eliminate potential product or process problems. Once the conceptual blocks of the design or process are known, there is no reason not to go ahead and create an FMEA and update on a regular basis as the product or process evolves.

T.3.2 Anytime

An FMEA can be created anytime during development or after development. Machine FMEAs can be done as part of a continuous improvement exercise.

T.4 Who Does the FMEA?

T.4.1 Engineers

Traditional FMEA development has typically fallen under the disciplines of electrical and mechanical engineering.

T.4.2   Designers

As one might suspect, the design FMEA (DFMEA) is a tool used by designers to anticipate failure modes introduced by poor or weak design approaches. Usually, a design team will create a DFMEA by listing the functional areas of the new (or old) product and then brainstorm potential failure modes for each function of the product. The DFMEA follows the normal structure of all FMEAs with the exception that the document is much less process-related and much more structure-related.

T.4.3   Social Practitioners

Nurses   A nurse could list the items currently afflicting a patient, possibly also including issues not seen yet. The nurse or nurse team would then consider potential failure modes; for example, a wrong dose or overdose of a medication for a specific patient.

Psychotherapists   A psychotherapist might consider potential pitfalls during sessions with a potentially violent client. By anticipating untoward behavior, the psychotherapist could head off any issues before they become significant.

Social workers   Social workers deal with many of the same issues as do psychotherapists. An FMEA would allow a social worker to anticipate issues, for example, in a chemical dependency clinic. Alternatively, the technique could be used in housing situations for the impoverished.

Lawyers   When in the courtroom, lawyers experience substantial amounts of give and take during testimony. A lawyer or a firm could anticipate witness responses with the use of a FMEA at any level of detail. Response would be prepared and practiced long before the testimony. Notes could be updated during the grand jury and the trial.

Human resources   Human resources staff could use the FMEA to prepare for employee dissatisfaction, for example, with a new compensation program. Additionally, the hiring process could be assessed with a process FMEA.

T.4.4   Anybody

Trips   Travelers can use the FMEA concept and apply to the process of taking a trip. What is a failure mode on a trip? Among these could be

- No hotel arrangements
- No car
- No contingency plans
- No phone numbers for help
- Inadequate fuel for vehicle

To some people, it may seem silly to go to a full FMEA for something as simple as a trip. A full J1739-style PFMEA is not necessary for any analysis of this type. Nobody will sit with a team and dream up numbers for occurrence on a trip. Severities are probably more like high, medium, low. We can create one of these with a very simple matrix (hand-drawn, spreadsheet, word processor, etc.) consisting of each item in the process, failure mode, cause, severity (if desired), detection (if desired), and a one or two line contingency plans to counteract the situation.

Projects A project is a process. We can analyze any process with a PFMEA. Some steps that can lead to a successful analysis are as follows:

1. Define the top-level steps or goals
2. Order these goals into time order (if possible)
3. For each goal, derive substeps or objectives
4. Order items again
5. If appropriate, define sub-substeps or targets
6. When the work breakdown is complete, begin to generate potential failure modes
7. At the end of the process, you should have a fairly good listing of potential problems, means of detection, and potential solutions (contingency plans)

In my experience, very few program managers or program teams take the time to do this kind of analysis. No project or program plan is ironclad, but consider the benefits of having a contingency plan for the breakdown of any part of the original or updated project.

Writing Good technical writing is the product of a writing plan (under most circumstances). If the technical writer or engineer has a plan, he or she also has a process. If he or she has a process, he or she can create a PFMEA for that process. What are some that issues might fall under consideration when writing?

- Research
- Tools
- Advanced layout
- Division of labor on a team effort
- Delivery

T.5 Where to Use the FMEA?

T.5.1 Manufacturing

The FMEA is used principally in manufacturing in the form of a process FMEA (PFMEA). In general, the PFMEA function numbers relate in database fashion to the work center numbers present in the line flow diagram and the process control plan. The manufacturing and industrial engineers try to anticipate issues that would affect the speed, cost, and quality of the manufactured product.

Manufacturing engineers may also use the machinery FMEA (MFMEA) to assess the potential for issues with the line equipment. The approach is generally very similar to the DFMEA because we normally assess structure rather than process, although a process approach could also be used.

T.5.2 Software Development

Software FMEAs are problematic. Failure modes in software become astronomical in number very quickly due to the number of potential paths through the software. One of us presented a method for using an FMEA for software to the SAE in 1998, wherein we suggested the FMEA as a tool for developing test cases. Currently, FMEA is not a good tool for software development because the FMEA is unlikely to add value due to the complexity of software problems.

T.5.3 Anywhere

The FMEA approach to anticipating problems is very simple and can be used for any structure or process. Its weakness lies in the subjectivity and qualitative nature of the approach. However, its simplicity helps to drive the use of the approach and any anticipation of or preparation for potential problems cannot be a bad thing. Figure T.1 shows a typical flow of activity that achieves a completed FMEA.

T.6 How Do We Build an FMEA?

T.6.1 FMEA Taxonomy

T.6.1.1 DFMEA

Item and function description In most FMEA formats, this item occurs in the very first column. It is as simple as defining the item under failure mode analysis. If we are developing an electronic or mechanical DFMEA, the item may refer to a drawing, a schematic, or a layout diagram. Entering the item alone is probably not sufficient: the "description" should also have a definition of the function of the item. Because it is the function that fails and not the item, we have to list each function separately from every other function while making it clear that they relate to the same item.

Cause If we can accept that failure modes are always the product of unplanned behavior at the outputs, then the causes come from one of two sources:

1. Unplanned behavior at an input,
2. Incorrect behavior at a transformation

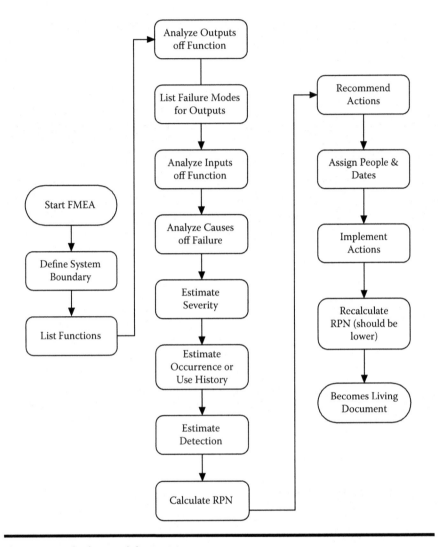

Figure T.1   The heart of the FMEA sequence.

When we are dealing strictly with hardware, we can expect that the bulk of the causes will fall into the class of input or inputs mated with an output or outputs. In some cases, however, we can treat the hardware as a form of program, where the input triggers a set of "internal" behaviors that, in turn, lead to an output. The collection of "internal behaviors" constitutes a transformation.

If we use the DFMEA to analyze software, every program is the relationship of inputs to outputs through some kind of transformation.

In the DFMEA layout, we should create a new "row" for each cause. This idea suggests that during the FMEA all untoward effects ultimately resolve down to one cause: a weakness of the FMEA method. If it is clear to us that multiple causes, independently or in interrelation, lead to a failure then we can use another tool—the fault tree—to analyze the failure mode. We do not want to use fault trees on every situation because they are labor intensive, typically much more so than the FMEA.

Severity The concept of severity in the DFMEA is significant for several reasons:

- We use it to calculate a "significance" using a combined value called risk priority number (RPN)
- We can designate items that present safety issues—which should receive analysis regardless of their RPN
- We establish a baseline against which we can compare our action results
- We start off with a list recommended by SAE J1739

I would like to note that the SAE J1739 list is not the final word on severity. MIL-STD-1629A, for example, uses the MIL-STD-882 (safety and hazard analysis) four category classification system to judge severity: Category I = Catastrophic; Category II = Critical; Category III = Marginal; and Category IV = Minor. The government recognizes that Categories I and II are significant regardless of their occurrence or detection and requires that these items be explicitly called out.

No matter what system the DFMEA team chooses to use, members of the team must all agree on the meaning of the various designations. SAE set up the J1739 format for a granularity of ten categories which is probably the most common arrangement. The DFMEA may elaborate on the definitions contained in the standard. If these documents are subject to customer review, all parties should agree on the meanings in the severity categories.

Classification The concept of "classification" is peculiar to the SAE J1739 and QS-9000 way of looking at DFMEA work. The DFMEA and/or design team can use the classification column to mark items that require process controls during manufacturing. If the team is using the DFMEA for software test case generation, this column makes no sense. Classification is not used in the calculation of the RPN value.

Occurrence In the DFMEA, "occurrence" relates to how often the failure mode occurs and uses a scale from one to ten. SAE J1739 recommends a set of ranking criteria in a relatively straightforward table. However, the DFMEA team can set any standard it wants; for example, in some cases the criteria for establishing occurrence are simply unavailable. In many cases, the team will not have empirical data to support its estimates, especially if it is working on a new product.

Another case occurs in a situation where the DFMEA becomes a tool for software. With a given software version and a given failure mode, the event will occur in all products that have that software. In most cases, it makes more sense to simply set the occurrence value at "5" and eliminate it from the calculation.

Design controls   For a design FMEA, design controls are typically one, some, or all of the following:

■ Reviews
■ Computer-aided tools
■ Testing
■ Inspection
■ Standards

The point is that we control designs by performing a collection of "best practices" which we believe result in better designs. In some cases— inspections, for example—we know empirically that the design control does in fact lead to a more defect-free design.

When the design control is a test, the FMEA team should call out the specific test document or test that is relevant to the particular failure mode. This way, the DFMEA becomes not only an anticipatory tool, but a means for specifying test cases. The related test document should show how the test cases flow from the DFMEA to the test description.

Detection description   The detection description or detection value provides a way to subjectively evaluate the capability of a design control to detect a defect in the product.

Risk priority number (RPN)   This value is the product of the severity, occurrence, and detection values determined after "actions taken":

$$RPN \ = \ severity \ \times \ occurrence \ \times \ detection$$

The higher the RPN, the more significant the failure mode. It is also important to remember criticality, which ties in most strongly with the idea of severity. For example, safety issues are significant regardless of the final RPN.

Recommended actions   In a design FMEA, recommended actions usually revolve around design modifications that lower the RPN value. It is also possible that the team may come up with no recommendations. Recommendations may also be procedural; that is, the problem may be so intractable that the team recommends the problem be handled in the instruction manual for the product or a data sheet.

Responsibility and target completion date   This column tries to establish ownership of issues as well as define a time at which the problem will come to resolution. Where there is no ownership, nobody does the design and detection work necessary to improve the product and the FMEA fails!

Where there is no completion date, we have no way to audit the FMEA to determine if the engineer ever took the recommended action. Again, the FMEA fails!

Actions taken   This column implies that either the FMEA team or the responsible engineer or engineers have taken the steps necessary to improve the product. If the "box" is empty, then, presumably no action has occurred and the product does not improve. This portion of the FMEA can also serve to record decisions to not act and point to an external document that defines that decision. Remember, the FMEA is a tool to help us work better, not a bureaucratic go-through-the-motions waste of time.

Severity, occurrence, and detection   After we take action, we recalculate the severity, occurrence, and detection values. All of the comments made in previous sections still apply. If we have done our job well, the new values should be decidedly lower than the old.

Final RPN   This value is the product of the new severity, occurrence, and detection values determined after "actions taken":

$$RPN = severity \times occurrence \times detection$$

T.6.1.2   PFMEA

Service   A service organization can capture "lessons learned" by recording events and responses in the PFMEA format. The service process can be assessed and problematic situations either addressed ahead of time or by updating the FMEA as needed.

Help desk   The IT help desk can use the FMEA format to record issues called in by customers/clients/employees. Responses can be maintained and updated as new issues occur. Even better would be the traditional anticipation of issues before they occur.

T.6.1.3   Policies

Procedures   The FMEA can be used to analyze a procedure because a procedure is really nothing but a formalized process. The same rules apply as would apply for any PFMEA.

Instructions   Instructions can be analyzed with either the structural or the process approach to the FMEA. Sometimes, instruction writers understand the work as they write the instruction because they are often subject matter experts, but a little anticipation using an FMEA would help with consideration of what the potential user is likely to experience.

T.6.1.4   Output Primacy

In my experience, most inadequate and unintelligible FMEAs I have seen resulted from starting on the wrong part of the document.

Outputs first When constructing any FMEA, but especially with design FMEAs, we always start with the outputs. A failure of an output is a failure mode. This choice has the benefits of

- Simplicity
- Consistency
- Sense
- Customer-oriented

Outputs are behaviors. Behaviors are what the customer (or the consumer of the activity) ultimately sees. Use the following thought experiment: Input fails, but output continues to function. Is there a failure mode? I say "no," because the customer/consumer does not see any failure. True, this event does not happen in the real world, but it does highlight the argument for output primacy.

The same rules apply for process FMEAs. Any given step in the process has an input and an output. The rule about outputs applies just the same.

Example: let's look at a speedometer. Typical failure modes for a gauge are the following:

- Pointer (needle) stuck in one position
- Pointer stuck at maximum value
- Pointer stuck at minimum value
- Pointer oscillating periodically
- Pointer oscillating aperiodically
- Pointer has different behavior in down direction than up direction

We can further subdivide the oscillation mal-behaviors into massive oscillations, small oscillations, etc.

Inputs are causes How can output be a "cause?" A cause is nearly always an input or combination of inputs which exhibit an undesirable action leading to the observable behavior we have called an "output."

Effect = sense-able Usually, we describe the effect by describing how the external entity observing the system will see it. For example, if the failure mode is a speedometer that always shows a zero value, the effect is that the driver does not know the speed of the vehicle, which, in turn, leads to safety issues.

In my experience, the people preparing an FMEA frequently confuse the effect with the output failure mode. The preparer must understand that the failure mode occurs at the output of the device under analysis, and the effect occurs in the observer of the device (the observer could be a sensor or a human being).

T.6.1.5   How Not to Do an FMEA

We indicate some ideas regarding poor FMEA practice:

- Have only one person build the entire document when a team is warranted,
- Create the FMEA after the product or process has already been developed,
- Only perform the FMEA because it is a PPAP requirement,
- Never review or update the FMEA once it has been created.

T.7   How Much Does It Cost to Do an FMEA?

T.7.1   Time

FMEAs for complex products and processes can take days or weeks to build. In cases like these, it might make sense to break the product or process into subsystems and defeat the problem "in detail." The only caveat when doing so is that we must now deal with the interfaces between the various subsystems.

T.7.2   Money

If the FMEA consumes resources in the form of time and people, then it also consumes money. However, we suggest that any potential lawsuit averted by anticipation of problem issues is a serious and worthy cost avoidance. Additionally, software that specifically supports the FMEA approach typically runs into thousands of dollars per seat, making it unlikely that all team members will have access to the product, particularly in smaller companies.

T.7.3   Team

The FMEA team should be cross-functional so that the approach does not stagnate on one discipline's particular mode of problem solving. In some cases, the "cross-pollination" of ideas leads to insights about potential behaviors of the product and makes for a better document.

T.7.4   How Much Does It Cost to Not Do an FMEA?

Potentially, the loss of millions of dollars in lawsuits or warranty can be avoided if the FMEA has been used as a serious tool for proactive elimination of issues. As always with quality-oriented tools, it is extremely difficult to quantify the cost benefit of doing an FMEA.

T.8 What Are the Strengths of the FMEA?

T.8.1 Anticipatory = Proactive

Because a properly created FMEA exists before the product or process becomes concrete, there is the possibility of averting problems *before* they happen. This consideration applies across the board, regardless of discipline or process.

T.8.2 Application

Any design Any design can be analyzed with an FMEA as long as a structure is discernible. This idea applies from very small designs like integrated circuits to very large designs like aircraft carriers and large buildings.

Any process As with design, any process can be analyzed with an FMEA, from very simple "kitchen" processes to complex social processes; for example, assessment of issues in a penal institution.

T.8.3 Simple Format

The FMEA format is very simple and it can be replicated very easily in a modern spreadsheet. The simplicity of the approach increases the likelihood that users will understand the process as well as actually create the document.

T.9 What Opportunities Do We Find with the FMEA?

T.9.1 Save Money

As we have noted above, it is very likely that cost can be avoided through the disciplined use of the FMEA. Eliminating "friction" will always lubricate the process.

T.9.2 Safety

Part of the FMEA severity assessment relates to the regulatory and safety-based behavior of the product or process. Not only do we wish to avoid injury to participants in the product or process, but again we avoid the costs associated with personal injury lawsuits.

T.9.3 Documentation

The FMEA provides a compact means of documenting

- Due diligence with respect to the product,
- "Lessons learned,"
- A learning tool for new employees.

T.9.4   Test Generation

The FMEA can be used to generate tests to *detect* problems with the product or process. It is essential here that the FMEA team not only indicate that detection is possible through testing, but that they also see to it that the test is explicitly specified in the FMEA, potentially using the capability of a spreadsheet to hyperlink to documents.

T.10   What Are the Weaknesses of the FMEA?

T.10.1   Loop Closing

If the FMEA never receives a review or audit after creation, a quality "loop" with the document remains open. Furthermore, the FMEA should be linked to testing and other evidence-oriented activities in order to keep the detection column "honest."

Auditing   FMEAs should be audited for completeness and for the seriousness of the approach. We have seen many FMEAs that were largely "hand waving" approaches to anticipatory problem solving. A good audit can reveal the level of thought that went into the document as well as help to force a revisitation to the document itself.

Detection   Tie the FMEA detection field to either laboratory testing or production test equipment testing. Based on experience, many FMEAs supposedly point to design verification or product validation testing, but do not provide any "hard closure" to prove that the detection did, in fact, occur.

T.10.2   Failure Mode Source

What caused the failure mode? Without good quantitative data, the team may *assume* it knows what causes a specific failure mode. Later experience may show otherwise. This situation is a byproduct of the subjective nature of the FMEA, which will only be as good as the team that creates and maintains it.

T.10.3   No Relationships

Simple failures   In general, FMEAs represent a collection of single-mode failures. It is extremely difficult to represent multimodal failures without a tree format to organize the information.

No fault trees   Fault trees can represent multimodal failures; the FMEA has no convenient representation for multimodal catastrophes.

T.10.4   Lack of Data

Occurrence   It is often difficult to find data associated with frequency of failure, particularly during the DFMEA. Suppliers do not necessarily make the failure rates of their products public knowledge. It is rarely seen on product data sheets. In some cases, the team will have to provide an estimate of the occurrence, perhaps setting it on the high side to force consideration of the failure mode during the action phase of the process.

Detection   Our experience suggests that teams are often optimistic about the ability of their supporting processes to detect failure modes. We have seen teams routinely throw in automotive generalities such as *design verification testing (DVT)* and *product validation testing (PVT)* without specifying how these test suites are supposed to detect the failure mode.

Severity   Severity is perhaps the easiest of the three components to consider, since the definitions are fairly clear. Our experience suggests that teams routinely *overestimate* the severity of the failure mode.

T.11   What Threats Does FMEA Pose?

T.11.1   Complacency

A truly complete FMEA can give the illusion that we have considered all the possibilities for failure. It is highly unlikely that we have done so. Additionally, we do not want to get to the point where we finish an FMEA, put it in the drawer, and forget about it. The FMEA remains a learning tool as well as a design assistant.

T.11.2   Automatism

Some software products exist that can take a bill of materials and generate a component-level DFMEA. Since this method involves little or no thought, it is unlikely that the team will learn anything from the document.

The only redeeming factor in an automatically generated FMEA lies in the ability to sort by risk priority number or severity and to pursue these items promptly. Additionally, the software will still leave open the recommended actions and following fields.

T.11.3   Failure to Close the Loop

If the results of warranty and returned merchandise are not fed back into the FMEA practice, the enterprise loses an opportunity to reinforce learning. FMEA-centric organizations can use the tool as a resource for capturing the following:

- Lessons learned,
- Warranty information,
- Design solutions garnered through experience,
- Production solutions garnered through experience,
- Equipment repair information.

The point is that the FMEA has many more uses than as a simple qualitative source document used to keep quality auditors at bay (at least with ISO/TS 16949:2002).

Chapter Notes

[1]MIL-STD-1629A, Military Standard Procedures for Performing a Failure Mode Effects and Criticality Analysis (Defense Acquisition University, 1983) p1.

Bibliography

[AIAGFMEA2001] Automotive Industry Action Group. *AIAG FMEA, AIAG Potential Failure Mode and Effects Analysis 3rd edition* Southfield, MI, 2001.

[AIAGAPQP1995] Automotive Industry Action Group. *AIAG APQP, AIAG Advanced Product Quality Planning and Control Plan 2nd edition* Southfield, MI, 1995.

[AIAGPPAP2006] Automotive Industry Action Group. *AIAG PPAP, AIAG Production Part Approval Process 4th edition* Southfield, MI, March 2006.

[AIAGMSA2002] Automotive Industry Action Group. *AIAG MSA, AIAG Measurement Systems Analysis 3rd edition* Southfield, MI, March 2002.

[AIAGQSA1998] Automotive Industry Action Group. *AIAG QSA, AIAG Quality Systems Assessment 2nd edition* Southfield, MI, March 1998.

[Bellman 2001] Bellman, G.M. *Getting Things Done When You Are Not in Charge.* New York: Fireside, 2001.

[Bridges 2001] Bridges, W. *The Way of Transition: Embracing Life's Most Difficult Moments.* Reading, MA: Perseus, 2001.

[Chawla and Renesch 1995] Chawla, S., and Renesch, J. *Learning Organizations: Developing Cultures for Tomorrow's Workplace* (1st ed.). Portland, OR: Productivity Press, 1995.

[CMMI 2003] Chrissis, Mary Beth, Konrad, Mike, and Shrum, Sandy, *CMMI Guidelines for Process Integration and Product Improvement.* Boston: Addison-Wesley, 2003.

[Cooke and Tate 2006] Cooke, Helen S. and Tate, Karen, *36 Hour Course in Project Management.* New York, NY: McGraw-Hill, 2006.

[Corley, Reed, Shedd, and Morehead 2002] Corley, R.N., Reed, O.L., Shedd, P.J., and Morehead, J.W. *The Legal and Regulatory Environment of Business* (12th ed.). Homewood, IL: Irwin/McGraw-Hill, 2002.

[DeCarlo 2004] DeCarlo, Doug, *Xtreme Project Management: Using Leadership, Principles, and Tools to Deliver Value in the Face of Volatility.* A Wiley Imprint, San Fransico, CA: Josey-Bass, 2004.

[Dobler and Burt 1995] Dobler, D. W. and Burt, D. N. *Purchasing and Supply Management: Text and Cases* (6th ed.). New York: McGraw-Hill, 1995.

[Eisenberg and Goodall 1997] Eisenberg, E. M. and Goodall, H. L. *Organizational Communication: Balancing Creativity and Constraint* (2nd ed.). New York, NY: St. Martin's Press, 1997.

[Fisher and Ury 1991] Fisher, R. and Ury, W. Getting to Yes: Negotiating Agreement Without Giving in (2nd ed.). New York: Penguin Books, 1991.

[Fleming and Koppelman 2000] Fleming, Q. W. and Koppelman, J. M. *Earned Value Project Management. Project Management Institute* (2nd ed.). Newtown Square, PA: PMI®, 2000.

[Gharajedaghi 1999] Gharajedaghi, J. *Systems Thinking: Managing Chaos and Complexity.* Massachusetts: Butterworth-Heinemann, 1999.

[Guffey 2003] Guffey, Mary Ellen. *Essentials of Business Communication* (6th ed.). Independence, KY: Thomson South-Western, 2003.

[Heldman 2005] Heldman, Kim. *Project Managers: Spotlight on Risk Management.* San Francisco, CA: Harbor Light Press, 2005.

[Heneman, Heneman, and Judge 1997] Heneman, H. G. III, Heneman, R. L., and Judge, T. A. *Staffing Organizations* (2nd ed.). Middleton, WI: Mendota House, Irwin, 1997.

[Hesselbein, Goldsmith, and Beckhard 1996] Hesselbein, F., Goldsmith, M., and Beckhard, R. *The Leader of the Future: New Visions, Strategies, and Practices for the Next Era.* The Drucker Foundation—Future Series. San Francisco, CA: Josey-Bass Publishers, 1996.

[Hill 2001] CH2M Hill Project Managers. *Project Delivery System: A System and Process for Benchmark Performance* (4th ed.). Denver, CO: Author, 2001.

[Hickman, Hickman, and Hickman 1992] Hickman, Thomas K., Hickman, William M. and Hickman, K. Thomas. *Global Purchasing: How To Buy Goods and Services in Foreign Markets.* Homewood, IL: Irwin Professional, 1992.

[IEEE1999] The Institute of Electrical and Electronics Engineers, Inc. *IEEE Std 1061-1998, IEEE Standard for a Software Quality Metrics Methodology* New York, NY, April 1999.

[IEEE1999] The Institute of Electrical and Electronics Engineers, Inc. *IEEE Std 1028-1997, IEEE Standard for Software Reviews* New York, NY, April 1999.

[Kaplan and Anderson 2007] Kaplan, Robert S., and Anderson, Steven R. *Time-Driven Activity-Based Costing: A Simpler and More Powerful Path to Higher Profits.* Boston: Harvard Business School Press, 2007.

[Kerzner 2001] Kerzner, H. Project Management: *A Systems Approach to Planning, Scheduling, and Controlling* (7th ed.) New York: John Wiley and Sons, 2001.

[Kim 1992a] Kim, D. H. *Systems Archetypes I.* Cambridge: Pegasus, 1992.

[Kim 1992b] Kim, D. H. *Toward Learning Organizations.* Cambridge: Pegasus, 1992.

[Kim 1994a] Kim, D. H. *Systems Archetypes II.* Cambridge: Pegasus, 1994.

[Kim 1994b] Kim, D. H. *Systems Thinking Tools.* Cambridge: Pegasus, 1994.

[Kruger 2005] Kruger, Gregory A. *A Statistician Looks at Inventory Management.* Milwaukee: American Society for Quality, Quality Progress, 2005.

[Lewis 2000] Lewis, J. P. *Project Planning, Scheduling and Control* (3rd ed.). New York: McGraw-Hill, 2000.

[Meredith and Mantel 2000] Meredith, J.R., and Mantel, S. J., Jr. *Project Management: A Managerial Approach* (4th ed.). New York: John Wiley and Sons, 2000.

[PMI2000] Project Management Institute. *A Guide to the Project Management Book of Knowledge (PMBOK® Guide 2000)* Newton Square, PA, 2000.

[MIL1980] Department of Defense. *Procedures for Performing a Failure Mode, Effects and Criticality Analysis* 24 November 1980.

[SAE1994] SAE International. *Potential Failure and Effects Analysis (Design FMEA) and Potential Failure Mode and Effects Analysis in Manufacturing and Assembly Processes (Process FMEA Reference Manual)* Warrendale, PA, July 1994.

[Senge 1994] Senge, P. *The Fifth Discipline: The Art and Practice of the Learning Organization.* New York: Doubleday-Currency, 1994.

[Senge, Roberts, Ross, Smith, and Kleiner 1994] Senge, P., Roberts, C., Ross, R.B., Smith, B.J., and Kleiner, A. *The Fifth Discipline Fieldbook: Strategies and Tools for Building a Learning Organization.* New York: Doubleday-Currency, 1994.

[Stepanek 2005] Stepanek, George. *Software Project Secrets. Why Software Projects Fail.* New York: Apress, 2005

[Thomsett 2002] Thomsett, Michael, C. *The Little Black Book of Project Management.* New York: AMACOM, 2002.

[Tracy 1996] Tracy, J. A. *The Fast Forward MBA in Finance.* New York: Wiley, 1996.

[Verzuh 1999] Verzuh, E. *The Fast Forward MBA in Project Management.* New York: John Wiley & Sons, 1999.

[Wideman 1992] Wideman, R. M. (Ed.) *Project & Program Risk Management: A Guide to Managing Project Risks and Opportunities.* Newtown, PA: Project Management Institute, 1992.

[Save International] *Valve Methodology Standard* (ed. 2007), Dayton, OH, p. 12.

[Shannon, 1948] Claude E. Shannon: A mathematical theory of communication, *Bell System Technical Journal*, Vol. 27, pp. 379–423, 623–656, 1948.

Index

Printed in the United States
by Baker & Taylor Publisher Services